Babak Bayani

Interacting Quantum-Dissipative Tunnelling Systems

Babak Bayani

Interacting Quantum-Dissipative Tunnelling Systems

Analytic derivation of the Caldeira-Leggett Hamiltonian from a microscopic model

Südwestdeutscher Verlag für Hochschulschriften

Impressum/Imprint (nur für Deutschland/only for Germany)
Bibliografische Information der Deutschen Nationalbibliothek: Die Deutsche Nationalbibliothek verzeichnet diese Publikation in der Deutschen Nationalbibliografie; detaillierte bibliografische Daten sind im Internet über http://dnb.d-nb.de abrufbar.
Alle in diesem Buch genannten Marken und Produktnamen unterliegen warenzeichen-, marken- oder patentrechtlichem Schutz bzw. sind Warenzeichen oder eingetragene Warenzeichen der jeweiligen Inhaber. Die Wiedergabe von Marken, Produktnamen, Gebrauchsnamen, Handelsnamen, Warenbezeichnungen u.s.w. in diesem Werk berechtigt auch ohne besondere Kennzeichnung nicht zu der Annahme, dass solche Namen im Sinne der Warenzeichen- und Markenschutzgesetzgebung als frei zu betrachten wären und daher von jedermann benutzt werden dürften.

Coverbild: www.ingimage.com

Verlag: Südwestdeutscher Verlag für Hochschulschriften GmbH & Co. KG
Heinrich-Böcking-Str. 6-8, 66121 Saarbrücken, Deutschland
Telefon +49 681 37 20 271-1, Telefax +49 681 37 20 271-0
Email: info@svh-verlag.de

Approved by: Mainz, U, Diss., 2012

Herstellung in Deutschland:
Schaltungsdienst Lange o.H.G., Berlin
Books on Demand GmbH, Norderstedt
Reha GmbH, Saarbrücken
Amazon Distribution GmbH, Leipzig
ISBN: 978-3-8381-3207-5

Imprint (only for USA, GB)
Bibliographic information published by the Deutsche Nationalbibliothek: The Deutsche Nationalbibliothek lists this publication in the Deutsche Nationalbibliografie; detailed bibliographic data are available in the Internet at http://dnb.d-nb.de.
Any brand names and product names mentioned in this book are subject to trademark, brand or patent protection and are trademarks or registered trademarks of their respective holders. The use of brand names, product names, common names, trade names, product descriptions etc. even without a particular marking in this works is in no way to be construed to mean that such names may be regarded as unrestricted in respect of trademark and brand protection legislation and could thus be used by anyone.

Cover image: www.ingimage.com

Publisher: Südwestdeutscher Verlag für Hochschulschriften GmbH & Co. KG
Heinrich-Böcking-Str. 6-8, 66121 Saarbrücken, Germany
Phone +49 681 37 20 271-1, Fax +49 681 37 20 271-0
Email: info@svh-verlag.de

Printed in the U.S.A.
Printed in the U.K. by (see last page)
ISBN: 978-3-8381-3207-5

Copyright © 2012 by the author and Südwestdeutscher Verlag für Hochschulschriften GmbH & Co. KG and licensors
All rights reserved. Saarbrücken 2012

Contents

1 Introduction   3

2 Preliminaries   11
   2.1 Mathematical Preliminaries   11
      2.1.1 Laplace Transform   11
      2.1.2 Inverse Laplace Transform   12
      2.1.3 Wick Rotation   12
      2.1.4 Feynman-Vernon Method   14
      2.1.5 Functions   15
   2.2 Physical Preliminaries   17
      2.2.1 Model   17
      2.2.2 Spectral Density   19
      2.2.3 Kernel   20
      2.2.4 Instantons   22
   2.3 The Caldeira-Leggett Model   24
   2.4 Summary   26

3 The 1-defect Model   27
   3.1 First Method   30
   3.2 Second Method   35
   3.3 Quantum Tunnelling   39

4 The 2-defect Model   43
   4.1 Two Defects   43
      4.1.1 Pre-diagonalisation Transformations   44
      4.1.2 Caldeira-Leggett form for $D \geq 2$ 45
      4.1.3 The Kernel for two anharmonic Bonds   47
   4.2 Tunnelling expectation value using extended NIBA   55
      4.2.1 Summary   74

5 Results and Conclusions   77

A Diagonalisation of the first approach   81

B Diagonalisation of the second approach   87

C Diagonalisation for two anharmonic bonds   93

D Calculation of the influence functional   103

E Density matrix for two anharmonic bonds   107

F Blip- and Sojourn charge summation   111

Chapter 1

Introduction

Quantum-mechanical effects such as tunnelling are experimentally well-verified on a microscopic scale. But to be a fundamental theory, quantum mechanics has to be valid for all scales. This problem was addressed by the famous thought experiment named Schroedinger's cat, proposed in 1935. In quantum mechanics linear combinations of solutions, labeled $|\Psi_+\rangle, |\Psi_-\rangle$, of the Schrödinger equation are also solutions. So not only $|\Psi_+\rangle, |\Psi_-\rangle$ but also $|\Psi\rangle = \lambda_+|\Psi_+\rangle + \lambda_-|\Psi_-\rangle$, with the normalised coefficients $|\lambda_+|^2 + |\lambda_-|^2 = 1$, solve the Schrödinger equation. Considering Schrödinger's thought experiment, how is it then possible, that a cat has only been observed as being dead **or** alive, but not in a superposition (dead **and** alive)? Or expressed more generally, when does a quantum system stop being a linear combination of states, each of which correspond to different states, and instead begin to have a unique classical description? First of all, one has to keep in mind, that the Schrödinger equation is valid only for isolated (quantum-)systems following a unitary evolution

$$i\hbar \frac{\partial}{\partial t}|\Psi\rangle = H|\Psi\rangle \tag{1.1}$$

In reality a totally isolated system does not exist. Only approximatively a system can been regarded as "isolated". Let an object have two possible initial states $|+\rangle, |-\rangle$ and a measuring device also having two states $|M_+\rangle, |M_-\rangle$. The measuring device is initially prepared in state $|M_-\rangle$ and reacts in the following way on the two possible states of the objects

$$(|+\rangle)|M_-\rangle \stackrel{\text{measurement}}{\rightarrow} |+\rangle|M_+\rangle \equiv |\Psi_{fin.}^+\rangle, \quad (|-\rangle)|M_-\rangle \stackrel{\text{measurement}}{\rightarrow} |-\rangle|M_-\rangle \equiv |\Psi_{fin.}^-\rangle \tag{1.2}$$

The measuring device acts as a pointer showing which state the object is in after the measurement. Now we would like to measure the state of the object. The initial state $|\Psi(t=0)\rangle \equiv |\Psi_{ini.}\rangle$ before the measurement is defined as

$$|\Psi_{ini.}\rangle = \left(\lambda_+|+\rangle + \lambda_-|-\rangle\right)|M_-\rangle \tag{1.3}$$

After performing the measurement using the rules defined in Eq. (1.2), one gets the final state $|\Psi(t>0)\rangle \equiv |\Psi_{fin.}\rangle$

$$|\Psi_{fin.}\rangle = \lambda_+|+\rangle|M_+\rangle + \lambda_-|-\rangle|M_-\rangle \equiv \lambda_+|\Psi_{fin.}^+\rangle + \lambda_-|\Psi_{fin.}^-\rangle \qquad (1.4)$$

Now the state $|\Psi_{fin.}\rangle$ is a superposition of states and does not show a definite result. Applied to Schrödinger's cat, the final state of the cat is a superposition of dead and alive. How can this paradox be resolved? Considering the state $|\Psi_{fin.}\rangle$ as a pure state, the density operator of such a pure state (1.4) is defined as

$$\rho_{pure} = |\Psi_{fin.}\rangle\langle\Psi_{fin.}| = \begin{pmatrix} |\lambda_+|^2 & \lambda_+\lambda_-^* \\ \lambda_+^*\lambda_- & |\lambda_-|^2 \end{pmatrix} \qquad (1.5)$$

, where the off-diagonal elements show interference between the components $|\Psi_{fin.}^+\rangle$ and $|\Psi_{fin.}^-\rangle$. The "observation" leads to the "collapse" of the wavefunction $|\Psi_{fin.}\rangle$ into state $|\Psi_{fin.}^+\rangle$ with the probability $|\lambda_+|^2$ or into state $|\Psi_{fin.}^-\rangle$ with the probability $|\lambda_-|^2 = 1 - |\lambda_+|^2$. The Copenhagen interpretation of quantum mechanics postulated this "collapse" of the wavefunction du to the "observation". This "collapse" occurs instantaneously and cannot be described by the Schrödinger equation, which follows a unitary evolution [1]. There are of course also other interpretations such as for example the "Many-worlds interpretation" by Everret [2], but this will not be discussed here. An interesting approach to the "collapse" of the wavefunction is the use of decoherence. As mentioned above, totally isolated systems do not exist, hence the Schrödinger equation describing the object coupled to the measurement device $|\Psi_{ini.}\rangle$ misses another term describing the effect of the environment $|E_U\rangle$. The environment consists of a very large number of states, basically all states of the whole universe except the already described cat and the measurement device. Expressed mathematically this reads

$$|E_U\rangle = |E_1\rangle|E_2\rangle\ldots|E_N\rangle, \quad N \text{ very large} \qquad (1.6)$$

A small deviation ϵ of one of the environmental states can be described as

$$\langle E_i'|E_i\rangle = 1 - \epsilon \qquad (1.7)$$

, where $|E_i'\rangle$ is the state that received the small deviation. Applying that for the very large number of environmental states yields

$$\langle E_U'|E_U\rangle = (1-\epsilon)^N \ll 1 \qquad (1.8)$$

This leads to the new form of Eq. (1.3)

$$|\Psi_{ini.}'\rangle = \left(\lambda_+|+\rangle + \lambda_-|-\rangle\right)|M_-\rangle|E_U\rangle \qquad (1.9)$$

, where $|E_U\rangle$ are the environmental states before the measurement. Using again Eq. (1.2) yields the new final state

$$|\Psi'_{fin.}\rangle = \lambda_+|+\rangle|M_+\rangle|E_{U+}\rangle + \lambda_-|-\rangle|M_-\rangle|E_{U-}\rangle \tag{1.10}$$

Tracing out the environmental states yields the reduced density operator describing the final state of the cat and measuring device

$$\begin{aligned}\rho_{red} &= Tr_{E_U}|\Psi'_{fin.}\rangle\langle\Psi'_{fin.}| = |\lambda_+|^2|\Psi^+_{fin.}\rangle\langle\Psi^+_{fin.}| + |\lambda_-|^2|\Psi^-_{fin.}\rangle\langle\Psi^-_{fin.}| \\ &+ \lambda_+\lambda_-^*|\Psi^+_{fin.}\rangle\langle\Psi^-_{fin.}|\underbrace{|E_{U+}\rangle\langle E_{U-}|}_{\approx 0} + \lambda_-\lambda_+^*|\Psi^-_{fin.}\rangle\langle\Psi^+_{fin.}|\underbrace{|E_{U-}\rangle\langle E_{U+}|}_{\approx 0}\end{aligned} \tag{1.11}$$

The reduced density operator describes an open system constantly interacting with the environment. Using the trace as in Eq. (1.11) is like averaging out the environmental degrees of freedom. With $\langle E_{U+}|E_{U-}\rangle = \langle E_{U-}|E_{U+}\rangle \approx 0$, which results from Eq. (1.8) the reduced density matrix defined in Eq. (1.11) yields

$$\rho_{red} \cong |\lambda_+|^2|\Psi_+\rangle\langle\Psi^+_{fin.}| + |\lambda_-|^2|\Psi^-_{fin.}\rangle\langle\Psi_-| = \begin{pmatrix} |\lambda_+|^2 & 0 \\ 0 & |\lambda_-|^2 \end{pmatrix} \tag{1.12}$$

The loss of quantum coherence to the environment leads to the possibility of describing quantum systems in the language of statistical mechanics.

The model calculated in this thesis consists of a quantum system coupled to an environment. The effect of tracing out the environment on the quantum system is considered in detail.

This thesis investigates the dissipative effect of the environment on a tunnelling two-state system (TSS) and on the interaction between tunnelling TSS. The low-temperature properties of amorphous materials have been attributed to the existence of tunnelling but noninteracting TSS in amorphous materials [3, 4]. Their central hypothesis is the statistical distribution of atoms (or groups of atoms) sitting more or less in TSS. There is no interaction between TSS considered. From that, they derive the linear specific heat, a universal property of amorphous materials and the anomalous thermal conductivity.

Later the interaction between those TSS in amorphous materials have been considered as the main reason for the observed low temperature anomalies [5]. Investigation of interacting TSS in amorphous solids have nowadays been widely investigated [6, 7]. These publications focus mainly on the low-temperature ($T < 100mK$) kinetics and thermodynamic properties of dielectric glasses. The interaction of the TSS is phenomenologically defined and arises from the strain field or the direct electrical dipole-dipole interaction with distance dependent strength decaying as R^{-3} in those systems [7]. Anomalous temperature behaviour in the relaxation properties at ultralow temperatures are found in those publications [6, 7].

This thesis is not interested in the phenomenologically derived TSS of dielectric glasses, but instead uses a microscopic model, where the position dependence of the TSS in the environment is investigated.

The model investigated in this thesis consists of a translationally invariant chain of N particles with harmonic nearest neighbour interaction with one exception. One, respectively two bonds are anharmonic. The anharmonicity is described by a symmetric double well potential in which the continuos degree of freedom can tunnel between the two minima.

One method of calculating such a tunnelling process is the instanton technique. Instantons were first applied in quantum chromodynamics (QCD) in the late '70s, early '80s [8, 9, 10, 11]. The instanton technique provided an exact finite-action solution to the classical Yang-Mills [8, 11] equations in Euclidean space-time. But also its use in statistical mechanics has been discussed extensively [12, 13].

Another possibility is to effectively restrict the anharmonic potential to the Hilbert space spanned by the two minima of the wells [14, 15, 16]. This allows a mapping onto the well known spin-boson Hamiltonian, which has been applied to many different physical systems some of which are discussed in [17, 18, 19, 20]. This mapping and the restrictions implied are discussed in detail in [14, 16].

Now we can investigate the dissipative effect the environment, sometimes named harmonic bath, has on the quantum mechanical tunnelling of the TSS. Quantum dissipation describes the quantum-mechanical analogon to the classical irreversible loss of energy. In quantum theory usually a Hamiltonian is used. The total energy of the full system is then a conserved quantity. A way to avoid this problem and being able to introduce dissipation is to split up the full system into two parts. The first is called the system, where dissipation occurs and the second is the environment, which receives the energy flowing out of the system. The energy is only transferred from one system to another and hence conserved.

The first approach of modelling such a system was done by Feynman and Vernon [21]. They modelled the environment as an infinite set of harmonic oscillators.

With the path integral formalism and for certain kinds of coupling to the quantum-mechanical system, the harmonic degrees of freedom can be eliminated, leaving a quantum-mechanical system showing dissipation.

In 1981 the idea of Feynman and Vernon has been applied to a specific system by Caldeira and Leggett [22]. Considering the magnetic flux trapped in a SQUID [23] and ignoring dissipation, a standard WKB[1] calculation shows quantum tunnelling as the dominant flux transition mechanism for temperatures $T \lesssim 100 mK$. Experiments [24, 25] with even higher temperatures $T \sim 1-2K$ have been interpreted as possible evidence for quantum tunnelling of the flux [22]. A SQUID (superconducting quantum interference device) is used to measure extremely weak magnetic fields. In Fig. 1.2 a dc-Squid (direct current) is shown. Its functionality is based

---

[1]Wentzel–Kramers–Brillouin

on the flux-quantisation $\phi_0 \equiv \frac{h}{q_s}$, ($h$ is the Planck constant and $q_s = 2e$ is the electron charge of the Cooper pairs) occuring in superconducting loops and the Josephson effect. Caldeira and Leggett mention a SQUID as a promising candidate to observe quantum tunnelling on a macroscopic scale [22]. The relevant macroscopic variable is the magnetic flux trapped in the superconducting loop.

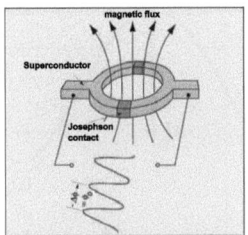

Figure 1.1: A schematic picture of a dc-SQUID (direct current superconducting quantum interference device). The Josephson junctions have to be thin enough to allow Cooper-pairs to tunnel through.

By considering linear coupling of the system with its environment, they find, that dissipation multiplies the tunnelling probability by a factor depending among other constants, on the phenomenological friction coefficient.

They discuss an imaginary time propagator for zero temperature (the derivation of those quantities follow, when the Caldeira-Leggett model is presented in the respective section). This propagator consists of an effective action, where the influence of the harmonic degrees of freedom on the tunnelling is described by a function[2] $\alpha(t - t')$. Caldeira and Leggett relate that function to the phenomenological friction coefficient η, by comparing the equations of motion of the phenomenological expression of the linear damped harmonic oscillator, with the ones achieved from the Caldeira-Leggett Lagrangian. Expressing the function $\alpha(t - t')$ in terms of the spectral density $J(\omega)$, they are able to provide the frequency dependence of this spectral density up to some cut-off frequency ω_c as $J(\omega \leq \omega_c) = \eta\omega$, in terms of the phenomenological friction coefficient η. The exact derivation is shown in the next chapter. With that, a connection of the quantum-mechanical effect of tunnelling to the classical effect of dissipation was made.

Model Hamiltonians of quantum systems coupled linearly to a bath of harmonic oscillators are well known nowadays as Caldeira-Leggett Hamiltonians and have been discussed in many articles e.g. [14, 16, 26, 27].

In this thesis the tunnelling of anharmonic bonds described by a symmetric double-well potential

---

[2]see section "The Caldeira-Leggett Model" for details

with the dissipative effect of the linearly coupled bath of harmonic oscillators, is investigated.

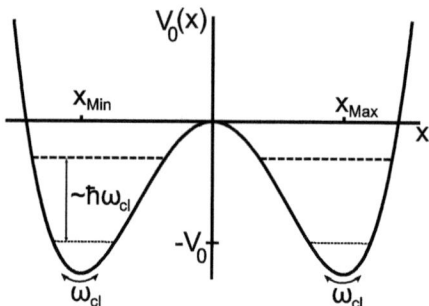

Figure 1.2: A double-well potential considered in the "two-state" limit. $\hbar\omega_{cl}$ denote the energy difference of the two-fold degenerate ground state to the first excited states. Symmetric TSS (regarded in this thesis) with two-fold degenerate ground states $x_{\text{Min}}, x_{\text{Max}}$ (without considering the coupling to the bath) of the system. V_0 represents the barrier height.

This potential is a TSS if the potential height $V_0 \gg \hbar\omega_{cl}$, where ω_{cl} is the classical small-oscillation frequency and the separation of the ground state from the first excited state is of order $\hbar\omega_{cl}$.

The effect of the environment on the tunnelling variable is described by the spectral density $J(\omega)$. The spectral density is derived and related to other quantities in chapter 2.2.1 - 2.2.3. The quantum dissipation generated by the harmonic bath depends qualitatively on the low-frequency behaviour of $J(\omega)$ [14, 16, 28]. The spectral density has a power-law form for frequencies much smaller than the cut-off frequency $J(\omega \leq \omega_c) \sim \omega^s$, where ω_c denotes the cut-off frequency introduced earlier. The exponent s classifies the dissipative influences. A detailed discussion on the effect of the spectral density will be given in section "The Caldeira-Leggett Model".

The indirect interaction of these TSS mediated through the harmonic bath depends on the microscopically derived spectral density $J(\omega)$. The harmonic bath leads to dissipative tunnelling behaviour of the TSS. This thesis is restricted to $T = 0$ and calculates the tunnelling probability of anharmonic bonds (described by TSS) in a chain of harmonic bonds (harmonic bath), whereas [6, 7] focus on the investigation of dielectric glasses and the anomalous temperature behaviour of their relaxation times.

Dubé and Stamp [29] investigate a similar system of interacting TSS as considered here. The main difference of the publication of Dubé and Stamp is allowing direct coupling and considering the continuous TSS coordinate of this thesis as a spin. The direct coupling leads to phenomena, that are not considered in this thesis. The calculations and approximations in their publications are not derived clearly and will be put on more stable ground in this thesis. From the microscopic model considered in this thesis the analytical derivation of the Caldeira-Leggett type Hamiltonian is performed. The Hamiltonian for this microscopic model is brought into the form of the Caldeira-Leggett Hamiltonian by separating the harmonic and anharmonic degrees of freedom analytically in different ways. First the position-dependence of one anharmonic bond is discussed. Since an open chain without periodic boundary conditions is considered, the position of the anharmonic bond effects the tunnelling behaviour. This will be shown in an analytic discussion.

Another motivation is the discussion of interacting TSS, since the model system allows to include more than one anharmonic bond. A finite distance between the anharmonic bonds describes indirect interaction between the TSS, whereas the limit of an infinite distance corresponds to the non-interacting case.

This thesis gives an insight in how interacting TSS change the tunnelling behaviour compared to the non-interacting case for the considered model system. In order to get the full physical effect of the interacting TSS, certain restrictions of the investigated system are considered. For one, both anharmonic bonds are located in the bulk, to diminish the effects an anharmonic bond receives at one of the boarders of the chain. Secondly finite distances D between both defects are considered, to get an indirect interaction of both anharmonic bonds via the harmonic bath.

This thesis splits up into 6 parts

- Chapter 1 gives an introduction to the topic of this thesis. It explains what has been done up to now on the topic of quantum dissipation and where quantum dissipation can be observed in experiment.

- Chapter 2 shows mathematical and physical preliminaries used throughout this thesis. This chapter is put at the beginning to allow more fluent reading later on, because the main mathematical and physical prescriptions are given. The phenomenological model of Caldeira and Leggett, used to calculate the effect of quantum dissipation, is presented and derived.

- Chapter 3 explains the used microscopic model in detail. A definition of all variables and the properties of the Hamiltonian are given. The reason for separating the harmonic from the anharmonic degrees of freedom will be explained.

    - Chapter 3.1 and 3.2 present two different ways of how to separate the harmonic from the anharmonic degrees of freedom. The first method is more intuitive but restricted to one dimensional problems, whereas the second method allows diagonalisation in every dimension. It will be shown that both methods are equivalent leading to the same Caldeira-Leggett Hamiltonian [22] (or Euclidean Lagrangian respectively).

– Chapter 3.3 discusses the position dependent quantum tunnelling of the one dimensional one anharmonic bond model system. Analytical results (within approximations) are given on how the position effects the tunnelling. Also a critical coupling constant, where tunnelling changes from oscillatory (de-localised state) to stochastic (localised state), is discussed.

– Chapter 3.4 introduces the notation and ways of calculating the probability and not the propagator (as in the sections before) of this microscopic system. The methods used here are equivalent to [14] and will be expanded later on for the two anharmonic bond case.

- Chapter 4 shows the calculation for the case of two anharmonic bonds interacting indirectly through the harmonic bath.

– Chapter 4.1 investigates how interaction between the TSS affect the tunnelling behaviour. First one has to perform a diagonalisation procedure as in the one anharmonic bond case. The technique of the first method is used for that.

– Chapter 4.2 generalizes the methods used in chapter 3.4 to the two anharmonic bond case. Here the techniques used in the one anharmonic bond case are used and explained and necessary additional simplifications are introduced. As a result one gets the expectation value of the two anharmonic bonds, tunnelling under the influence of the harmonic bath, which introduces dissipative effects. The dependency on the choice of initial and final state is also presented.

- Chapter 5 concludes and summarises the results achieved in this thesis.

- Chapter 6 is the Appendix, where certain procedures, such as diagonalisation of the different methods, derivation of the probability using the Feynman-Vernon path integral technique, are discussed in detail.

Chapter 2

Preliminaries

2.1 Mathematical Preliminaries

In this chapter mathematical definitions and notations are introduced, which are essential for this thesis. By introducing the mathematical notations and definitions here, the Appendix will be restricted to calculations, which are, due to their length, left out of the main part of the thesis. This will hopefully give the reader a well organised presentation of this thesis and allow fluent reading.

2.1.1 Laplace Transform

Let $f(t)$ be a function, then $\tilde{f}(\lambda)$ is the Laplace transform of this function $f : [0, \infty[\to \mathbb{C}$, defined by

$$\tilde{f}(\lambda) = \mathcal{L}\{f(t)\} = \int_0^\infty dt\, e^{-\lambda t} f(t), \qquad \lambda \in \mathbb{C},\, \Re(\lambda) > 0 \tag{2.1}$$

, where f must be locally integrable on $[0, \infty[$. The Laplace transform will be used in this thesis to transform convolutions into a product of Laplace transforms

$$\mathcal{L}\{f * g\} = \mathcal{L}\left\{\int_0^t dt'\, f(t')g(t-t')\right\} = \mathcal{L}\{f(t)\}\mathcal{L}\{g(t)\} = \tilde{f}(\lambda)\tilde{g}(\lambda) \tag{2.2}$$

The above equation can be proven as follows. Let $\tilde{f}(\lambda)\tilde{g}(\lambda) = \tilde{h}(\lambda)$, then

$$\tilde{h}(\lambda) = \int_0^\infty dt\, e^{-\lambda t} \left(\int_0^t dt'\, f(t')g(t-t')\right)$$

$$\tilde{h}(\lambda) = \int_0^\infty dt \int_0^t dt'\, e^{-\lambda t} f(t')g(t-t') \tag{2.3}$$

The first integration is carried out from $0 \leq t' \leq t$. Changing the order of integration results in

$$\tilde{h}(\lambda) = \int_0^\infty dt' \int_{t'}^\infty dt\, e^{-\lambda t} f(t')g(t-t') \tag{2.4}$$

Now changing the variable t to $t'' = t - t'$; $dt'' = dt$, the region of integration becomes $t' \geq 0, t'' \geq 0$

$$\begin{aligned}\tilde{h}(\lambda) &= \int_0^\infty dt' \int_0^\infty dt''\, e^{-\lambda(t'+t'')} f(t')g(t'') \tag{2.5}\\ &= \left(\int_0^\infty dt'\, e^{-\lambda t'} f(t')\right)\left(\int_0^\infty dt''\, e^{-\lambda t''} g(t'')\right) \\ &= \tilde{f}(\lambda)\tilde{g}(\lambda) \tag{2.6}\end{aligned}$$

2.1.2 Inverse Laplace Transform

The inverse Laplace transform is defined as

$$f(t) = \mathcal{L}^{-1}\{\tilde{f}(\lambda)\} = \frac{1}{2\pi i}\int_C d\lambda\, e^{\lambda t} \tilde{f}(\lambda) \tag{2.7}$$

, where C denotes the standard Bromwich contour, i.e. any contour in the complex plane from $-i\infty$ to $i\infty$ lying entirely to the right of all singularities of $\tilde{f}(\lambda)$. In this thesis, the inverse Laplace transform is not of primary interest. Mainly the Laplace transform will be used to discuss the poles of a Laplace transform $\tilde{f}(\lambda)$ to get information about the function in time space $f(t)$.

2.1.3 Wick Rotation

Wick rotations are commonly used to connect statistical mechanics with quantum mechanics by replacing the inverse temperature $\frac{1}{k_B T}$ with the imaginary time $\frac{it}{\hbar}$. In this thesis the Wick rotation is needed to achieve the transition amplitude from the instanton solutions of the path integral formulation. The propagator in real time reads

$$G(x_f, t; x_i, t_0) = \langle x_f | e^{-\frac{iH(t-t_0)}{\hbar}} | x_i \rangle = \int_{x(t_0)=x_i}^{x(t)=x_f} \mathcal{D}x(t)\, e^{\frac{iS[x(t)]}{\hbar}} \tag{2.8}$$

Choosing the starting time $t_0 = 0$ and performing the Wick rotation

$$t = -i\tau \tag{2.9}$$

2.1. MATHEMATICAL PRELIMINARIES

yields

$$G_E(x_f, \tau = T; x_i, \tau = 0) = \langle x_f | e^{-\frac{H\tau}{\hbar}} | x_i \rangle = \int_{x_i}^{x_f} \mathcal{D}x(\tau) \, e^{-\frac{S^E[x(\tau)]}{\hbar}} \tag{2.10}$$

, where the index E stands for Euclidean and indicates, that the Wick rotation has been performed.
In this thesis the Euclidean action is defined for a free particle in a double well potential

$$S^E = \int d\tau \left(\frac{m}{2} \left(\frac{dx}{d\tau} \right)^2 + V(x) \right) \tag{2.11}$$

, which is nothing but the normal action with a sign change in the potential $V(x) \to -V(x)$. The advantage can be easily seen in the following example:
Lets consider a symmetric double well potential

$$V(x) = x^4 + \sigma x^2, \qquad \sigma < 0 \tag{2.12}$$

where we want to calculate the tunnelling amplitude $G(x_f, t; x_i, 0)$. Classically the particle sitting in one of the two wells has no possible way to reach the other well for energies smaller than the barrier. In other words the equations of motion resulting from the Lagrangian has no solutions. By performing the Wick rotation and hence changing the sign of the potential, the particle is able to tunnel from one well to the other. That means the Euclidean Lagrangian has a solution. This solution of the Euclidean integral is named "kink"-solution and is an example of an instanton solution. The name results from the following fact. By changing the sign of the potential as discussed before the wells become hills and the particle rolls from one *hill* to the other. For the double well potential introduced above the solution of the Euclidean equations of motion is a hyperbolic tangent. The shape of this solutions lead to the name instanton solution, because the hyperbolic tangent stays infinitely long at -1 and then as the argument approaches zero it *instantaneously* flips to $+1$. *Instantaneously* is not meant as a sharp step or a discontinuity, but as a comparison of a fast change in a short period of the argument around zero, compared to the almost not changing value of the hyperbolic tangent for the rest of the arguments value.

2.1.4 Feynman-Vernon Method

Here the Feynman-Vernon method of describing quantum dissipation via a system coupled linearly to a harmonic bath, is used. Let there be a Hamiltonian of the following form

$$\begin{aligned} H &= H_{\text{bath}} + H_{\text{sys.}} + H_I \\ H_{\text{bath}} &= H_{\text{bath}}(\{P_\sigma\}, \{Q_\sigma\}) \\ H_{\text{sys.}} &= H_{\text{sys.}}(p, q) \\ H_I &= q \sum_\sigma c_\sigma Q_\sigma \end{aligned} \quad (2.13)$$

, where c_σ are the coupling constants of the bath coordinates $\{Q_\sigma\}$ to the system coordinate q. Using the Liouville-von-Neumann equation for the time dependent density operator $\rho_{tot}(t)$ of the total Hamiltonian, we can write

$$\frac{d}{dt}\rho_{tot}(t) = -\frac{i}{\hbar}[H, \rho_{tot}(t)] \;, \qquad \rho_{tot}(t) = e^{-\frac{i}{\hbar}Ht}\rho_{tot}(0)e^{\frac{i}{\hbar}Ht} \quad (2.14)$$

The full density matrix element in spatial representation reads

$$\begin{aligned} \langle\{Q_\sigma\}, q|\rho_{tot}(t)|q', \{Q'_\sigma\}\rangle = \int &dq^0 \, dq'^0 \, d\{Q^0_\sigma\} \, d\{Q'^0_\sigma\} \, \langle\{Q_\sigma\}, q|e^{-\frac{i}{\hbar}Ht}|q^0, \{Q^0_\sigma\}\rangle \\ &\cdot \langle\{Q^0_\sigma\}, q^0|\rho_{tot}(0)|q'^0, \{Q'^0_\sigma\}\rangle \\ &\cdot \langle\{Q'^0_\sigma\}, q'^0|e^{\frac{i}{\hbar}Ht}|q', \{Q'_\sigma\}\rangle \end{aligned} \quad (2.15)$$

Let an operator $\hat{A}$ have a matrix representation $A_{mn} = \langle m|\hat{A}|n\rangle$, then the trace is defined as

$$Tr\,\hat{A} = \sum_n \langle n|\hat{A}|n\rangle = \sum_n A_{nn} \quad (2.16)$$

Now in the case considered here the density operator $\rho_{tot}(t)$ includes the behaviour of the bath and the system. Tracing out the bath degrees of freedom $\{Q_\sigma\}, \{Q'_\sigma\}$ in the way shown above, we get

$$\begin{aligned} \rho_{red}(t) &= Tr_{bath}\,\rho_{tot}(t) \\ \Rightarrow \langle q|\rho_{red}(t)|q'\rangle &= \int dq^0 \, dq'^0 \, d\{Q^0_\sigma\} \, d\{Q'^0_\sigma\} \, d\{Q_\sigma\} \, \langle\{Q_\sigma\}, q|e^{-\frac{i}{\hbar}Ht}|q^0, \{Q^0_\sigma\}\rangle \\ &\quad \cdot \langle\{Q^0_\sigma\}, q^0|\rho_{tot}(0)|q'^0, \{Q'^0_\sigma\}\rangle \langle\{Q'^0_\sigma\}, q'^0|e^{\frac{i}{\hbar}Ht}|q', \{Q_\sigma\}\rangle \end{aligned} \quad (2.17)$$

Assuming the density matrix has factorising initial conditions $\rho_{tot}(0) = \rho_{red}(0) \otimes \rho_{bath}$, we are able to write down the final result for the propagator matrix elements

$$\langle q|\rho_{red}(t)|q'\rangle = \int dq^0 \, dq'^0 \, \langle q^0|\rho_{red}(0)|q'^0\rangle \int \mathcal{D}q \int \mathcal{D}q' \, e^{\frac{i}{\hbar}(S_{sys}[q] - S_{sys}[q'])} \mathcal{F}[q, q'] \quad (2.18)$$

The last term $\mathcal{F}$ is called the influence functional in literature and describes the effect of the bath on the system. For zero system-bath coupling the influence functional yields one. This influence functional is derived in greater detail for the more general case of two system coordinates coupling to a harmonic bath in Appendix D.

2.1. MATHEMATICAL PRELIMINARIES

2.1.5 Functions

In this subsection, all functions used throughout this thesis will be defined. First the generating functions for canonical transformations are introduced. They are defined in the following way

$$
\begin{aligned}
p(q,q',t) &= \frac{\partial R_1}{\partial q}(q,q',t), & p'(q,q',t) &= -\frac{\partial R_1}{\partial q'}(q,q',t) \\
p(q,p',t) &= \frac{\partial R_2}{\partial q}(q,p',t), & q'(q,p',t) &= \frac{\partial R_2}{\partial p'}(q,p',t) \\
p'(p,q',t) &= -\frac{\partial R_3}{\partial q'}(p,q',t), & q(p,q',t) &= -\frac{\partial R_3}{\partial p}(p,q',t) \\
q(p,p',t) &= -\frac{\partial R_4}{\partial p}(p,p',t), & q'(p,p',t) &= \frac{\partial R_4}{\partial p'}(p,p',t)
\end{aligned}
\tag{2.19}
$$

A derivation of the generating functions, as a special kind of point transformation, will not be given here, but can be seen in [30].

The next definition is for the propagator or Green's function. A Green's function is a function used to to solve an inhomogeneous differential equation subject to boundary conditions. Let L be a linear differential operator, f be the inhomogeneity and y the function we would like to find a solution for, then the following equation

$$Ly(x) = f(x) \tag{2.20}$$

can be solved by a Green's function $G(x)$ with the following property

$$LG(x) = \delta(x) \tag{2.21}$$

The solution is then

$$y(x) = (G * f)(x) = \int f(x_0) G(x - x_0)\, dx_0 \tag{2.22}$$

This can be seen by applying the definition of Eq. (2.20), as shown below

$$
\begin{aligned}
Ly(x) &= f(x) \\
\Leftrightarrow L\left(\int f(x_0) G(x - x_0) dx_0\right) &= f(x) \\
\Leftrightarrow \int f(x_0) LG(x - x_0) &= f(x) \\
\Leftrightarrow \int f(x_0) \delta(x - x_0) &= f(x) \\
\Leftrightarrow f(x) &= f(x)
\end{aligned}
\tag{2.23}
$$

The functions Si(x), Ci(x) are Sine and Cosine Integrals, defined as

$$\text{Si}(x) = \int_0^x dt\, \frac{\sin(t)}{t}$$

$$\text{Ci}(x) = \gamma + \ln(x) + \int_0^x dt\, \frac{\cos(t) - 1}{t} \tag{2.24}$$

, where γ is the Euler-Mascheroni constant, defined as

$$\gamma = \lim_{n \to \infty} \left(\sum_{k=1}^n \frac{1}{k} - \ln(n) \right) \tag{2.25}$$

2.2 Physical Preliminaries

In this chapter functions needed for an essential understanding of this thesis are introduced. These functions will be presented, explained and their use will be shown shortly in this section in order to leave out the information and improve easy reading in the main part. As in the section "Mathematical Preliminaries", the reason for this section is to give the reader a well organised presentation of this thesis and allow fluent reading.

2.2.1 Model

In order to investigate quantum dissipation, a simple model Hamiltonian (or equivalently a Euclidean Lagrangian [22]), known as the Caldeira-Leggett model, is introduced.

$$H = H_{bath} + H_{sys} + H_{int} \quad (2.26)$$
$$H_{bath} = \frac{1}{2}\sum_{\sigma=1}^{N}\left(\frac{P_\sigma^2}{m_\sigma} + m_\sigma \omega_\sigma^2 Q_\sigma^2\right)$$
$$H_{sys} = \frac{p^2}{2m} + V(q)$$
$$H_{int} = -q\sum_{\sigma=1}^{N} c_\sigma Q_\sigma + \Delta V(q)$$

q, p are the coordinate and momentum and $V(q)$ is the potential of the system. $\Delta V(q)$ is a counter term, which depends on the parameters m_σ, ω_σ, only [16] (chapter 3). Its physical significance is seen in Eqs. (2.36), (2.37). Those parameters are the masses and the frequencies of the harmonic bath coordinates, respectively. Q_σ, P_σ are the coordinate and momentum of the harmonic bath, where the index σ denotes the individual bath modes running from 1 to N. c_σ is the coupling constant. In this model we used linear coupling of the system to the bath. Other types of coupling are possible, but are not used throughout this thesis.

If we want to solely describe dissipation with our model without renormalising the potential $V(q)$, the counter term must have the following form [16]

$$\Delta V(q) = \sum_{\sigma=1}^{N} \frac{c_\sigma^2}{2m_\sigma \omega_\sigma^2} q^2 \quad (2.27)$$

Including the above counter term, we are able to write the Hamiltonian in a different form

$$H = \frac{p^2}{2m} + V(q) + \frac{1}{2}\sum_{\sigma=1}^{N}\left[\frac{P_\sigma^2}{m_\sigma} + m_\sigma \omega_\sigma^2 \left(Q_\sigma - \frac{c_\sigma}{m_\sigma \omega_\sigma^2} q\right)^2\right] \quad (2.28)$$

The equations of motions from a Hamiltonian are easily achieved

$$\dot{p} = -\frac{\partial H}{\partial q}$$
$$\dot{q} = \frac{\partial H}{\partial p} \qquad (2.29)$$

they read for the Hamiltonian of Eq. (2.28)

$$m\ddot{q} + \frac{\partial V(q)}{\partial q} + \sum_{\sigma=1}^{N} \frac{c_\sigma^2}{m_\sigma \omega_\sigma^2} q = \sum_{\sigma=1}^{N} c_\sigma Q_\sigma$$
$$m_\sigma \ddot{Q}_\sigma + m_\sigma \omega_\sigma^2 Q_\sigma = c_\sigma q \qquad (2.30)$$

The solution for $Q_\sigma(t)$ can be achieved by Green's functions techniques introduced earlier in section "Mathematical Preliminaries". They are

$$Q_\sigma(t) = Q_\sigma(0)\cos(\omega_\sigma t) + \frac{P_\sigma(0)}{m_\sigma \omega_\sigma}\sin(\omega_\sigma t) + \frac{c_\sigma}{m_\sigma \omega_\sigma}\int_0^t dt'\sin(\omega_\sigma[t-t'])q(t') \qquad (2.31)$$

Now following the notation and technique of [16], we get by integration by parts

$$Q_\sigma(t) = Q_\sigma(0)\cos(\omega_\sigma t) + \frac{P_\sigma(0)}{m_\sigma \omega_\sigma}\sin(\omega_\sigma t)$$
$$+ \frac{c_\sigma}{m_\sigma \omega_\sigma^2}\left(q(t) - q(0)\cos(\omega_\sigma t) - \int_0^t dt'\cos(\omega_\sigma[t-t'])\dot{q}(t')\right) \qquad (2.32)$$

Now using this solution and plugging it into Eq. (2.30), we get

$$m\ddot{q}(t) + m\int_0^t dt'\gamma(t-t')\dot{q}(t') + \frac{\partial V(q)}{\partial q} = -m\gamma(t)q(0) + \zeta(t) \qquad (2.33)$$

, with the force

$$\zeta(t) = \sum_\sigma c_\sigma\left(Q_\sigma(0)\cos(\omega_\sigma t) + \frac{P_\sigma(0)}{m_\sigma \omega_\sigma}\sin(\omega_\sigma t)\right) \qquad (2.34)$$

and the memory-friction kernel obeying causality ($\gamma(t) = 0$ for $t < 0$)

$$\gamma(t-t') = \frac{\Theta(t-t')}{m}\sum_\sigma \frac{c_\sigma^2}{m_\sigma \omega_\sigma^2}\cos(\omega_\sigma[t-t']) \qquad (2.35)$$

The Eq. (2.33) is a Langevin-type equation with an additional term $-m\gamma(t)q(0)$ depending on the initial value $q(0)$. This additional term can be included in the random force by the following definition

$$\xi(t) = \zeta(t) - m\gamma(t)q(0) \qquad (2.36)$$

2.2. PHYSICAL PRELIMINARIES

The properties of a classical Langevin equation are well known. Taking the average of the initial values with respect to the shifted canonical equilibrium density

$$\rho_{bath} = Z^{-1} e^{-\beta \sum_{\sigma=1}^{N}\left(\frac{P_\sigma^2(0)}{2m_\sigma} + \frac{m_\sigma \omega_\sigma^2}{2}\left(Q_\sigma(0) - \frac{c_\sigma}{m_\sigma \omega_\sigma^2}q(0)\right)^2\right)} \tag{2.37}$$

$\xi(t)$ becomes a fluctuating force with Gaussian statistical properties

$$\begin{aligned}\langle \xi(t) \rangle_{\rho_{bath}} &= 0 \\ \langle \xi(t)\xi(t') \rangle_{\rho_{bath}} &= \frac{m}{\beta}\gamma(t-t')\end{aligned} \tag{2.38}$$

, where $\beta = \frac{1}{k_B T}$.

Fourier transforming the memory-friction kernel of Eq. (2.35), we get

$$\tilde{\gamma}(\omega) = -\frac{i\omega}{m}\sum_{\sigma=1}^{N}\frac{c_\sigma^2}{m_\sigma \omega_\sigma^2}\lim_{\epsilon \to 0^+}\frac{1}{\omega_\sigma^2 - \omega^2 - i\epsilon\omega}$$

2.2.2 Spectral Density

The spectral density function $J(\omega)$, contains the complete information about the effect of the environment. It is defined as

$$J(\omega) = \frac{\pi}{2}\sum_{\sigma=1}^{N}\frac{c_\sigma^2}{m_\sigma \omega_\sigma}\delta(\omega - \omega_\sigma) \tag{2.39}$$

Considering the spectral density as a smooth function of ω and performing the thermodynamic limit $N \to \infty$, we are able to rewrite the Fourier Transform of the memory-friction kernel (2.39) in terms of the spectral density [16]

$$\tilde{\gamma}(\omega) = \lim_{\epsilon \to 0^+} -\frac{i\omega}{m}\frac{2}{\pi}\int_0^\infty d\omega' \frac{J(\omega')}{\omega'}\frac{1}{\omega'^2 - \omega^2 - i\epsilon\omega} \tag{2.40}$$

Up to now, this spectral density has only been derived phenomenologically. In those derivations [14, 16, 22] the following assumptions were made. $J(\omega)$ is considered a reasonably smooth function of ω and that it is of the form ω^s, $s > 0$ up to some cut-off frequency ω_c.

In this thesis the spectral density will be derived analytically for a microscopic model in the thermodynamic limit ($N \to \infty$). The variables $c_\sigma, m_\sigma, \omega_\sigma$ can be calculated from the microscopic model, which is shown for the case of one and two anharmonic bonds.

The spectral density is of main importance for this thesis and is derived showing the behaviour $\sim \omega^s$ assumed by Leggett et al.. Three different cases occur generally for $J(\omega) \sim \omega^s$:

- $0 < s < 1$ the sub-ohmic case

- $s = 1$ the ohmic case

- $1 < s$ the super-ohmic case

The sub-ohmic case will not be discussed in this thesis. It has been shown [14], that sub-ohmic dissipation leads to complete localisation at $T = 0$, whereas the super-ohmic case yields weakly damped oscillations [14]. The critical dimension is achieved for $s = 1$, the ohmic case, where the coupling constant $c_\sigma \sim C$ of the anharmonic bond coordinate to the harmonic bath coordinates, has a critical value yielding tunnelling for $C < C_{crit.}$ and localisation for the other case.

2.2.3 Kernel

But the spectral density is not only useful to express the Fourier Transform of the memory-friction kernel $\tilde{\gamma}(\omega)$, it can also be used in a different approach of describing quantum dissipation.

In [22] the authors calculate the Euclidean propagator

$$G_E(q_f, \{Q_\sigma^{(f)}\}, T; q_i, \{Q_\sigma^{(i)}\}, 0) = \int_{q_i}^{q_f} \mathcal{D}q(\tau) \int_{\{Q_\sigma^{(i)}\}}^{\{Q_\sigma^{(f)}\}} \mathcal{D}\{Q_\sigma(\tau)\} e^{\underbrace{-\frac{1}{\hbar}\int_0^T d\tau' L^E(q(\tau'), \dot{q}(\tau'); \{Q_\sigma(\tau')\}, \{\dot{Q}_\sigma(\tau')\})}_{s_E}}$$

using the Euclidean Lagrangian (which is nothing but the Legendre Transform of the Hamiltonian (2.28) in Euclidean form)

$$L^E = \frac{m}{2}\dot{q}^2(\tau) + V(q) + \sum_{\sigma=1}^{N} \frac{m_\sigma}{2}\left[\dot{Q}_\sigma^2(\tau) + \omega_\sigma^2\left(Q_\sigma(\tau) - \frac{c_\sigma}{m_\sigma \omega_\sigma^2}q(\tau)\right)^2\right] \quad (2.41)$$

To calculate the Euclidean tunnelling propagator, the Euclidean action S_E can be split up into two parts

$$S^E = S_0^E + S_{harm,int}^E$$

, where the integral of the first two terms are S_0^E without interaction of the environment, whereas the interaction with the environment is fully captured in $S_{harm,int}^E$. The discussion of this part is done explicitly in the next subsection "Instantons".

Now the elimination of the harmonic degrees of freedom is performed. The paths $q(\tau), \{Q_\sigma(\tau)\}$ are periodically continued outside the range of $0 \leq \tau < T$ (where τ denotes the imaginary *time*

2.2. PHYSICAL PRELIMINARIES

variable) by writing them as a Fourier series [16]

$$q(\tau) = \frac{1}{T} \sum_{n=-\infty}^{\infty} q_n e^{i\nu_n \tau}$$

$$Q_\sigma(\tau) = \frac{1}{T} \sum_{n=-\infty}^{\infty} Q_{\sigma,n} e^{i\nu_n \tau} \tag{2.42}$$

where $\nu_n = 2\pi n/T$ is the frequency of the Fourier series. Applying this transformation to the Euclidean action yields

$$S^E_{harm,int} = \frac{1}{T} \sum_\sigma \sum_{n=-\infty}^{\infty} \frac{m_\sigma}{2} \left(\nu_n^2 |Q_{\sigma,n}|^2 + \omega_\sigma^2 \left| Q_{\sigma,n} - \frac{c_\sigma}{m_\sigma \omega_\sigma^2} q_n \right|^2 \right)$$

Next, $Q_{\sigma,n}$ will be decomposed into a classical term $\overline{Q}_{\sigma,n}$ and a deviation $y_{\sigma,n}$ describing quantum fluctuations [16]

$$Q_{\sigma,n} = \overline{Q}_{\sigma,n} + y_{\sigma,n} = \frac{c_\sigma}{m_\sigma(\nu_n^2 + \omega_\sigma^2)} q_n + y_{\sigma,n} \tag{2.43}$$

This result is trivially achieved using Eqs. (2.42) and (2.30). In the second part the classical solutions of the Euclidean equations of motion are used. Since $\overline{Q}_\sigma(\tau)$ is a stationary point of the action, the term linear in the deviation $y_{\sigma,n}$ is eliminated. With this approach it is possible to decouple the bilinear forms containing the anharmonic bond and the harmonic degrees of freedom

$$S^E_{harm,int} = S^E_{harm} + S^E_{infl.}$$

$$S^E_{harm} = \frac{1}{T} \sum_\sigma \sum_{n=-\infty}^{\infty} \frac{m_\sigma}{2} (\nu_n^2 + \omega_\sigma^2) |y_{\sigma,n}|^2 = \sum_\sigma \int_0^T d\tau \frac{m_\sigma}{2} \left(\dot{y}_{\sigma,n}^2 + \omega_\sigma^2 y_{\sigma,n}^2 \right)$$

$$S^E_{infl.} = \frac{1}{T} \sum_\sigma \frac{c_\sigma^2}{2m_\sigma} \sum_{n=-\infty}^{\infty} \left(\frac{|q_{M,n}|^2}{\omega_\sigma^2} - \frac{|q_{M,n}|^2}{\nu_n^2 + \omega_\sigma^2} \right) \tag{2.44}$$

The first term in $S^E_{infl.}$ originates from the potential counter term $\frac{C}{4} q_n^2$. Changing to the time representation the influence action (2.44) reads [16]

$$S^E_{infl.} = \int_0^T d\tau \int_0^\tau d\tau' \, k(\tau - \tau') q(\tau) q(\tau')$$

$$k(\tau) = \frac{1}{T} \sum_\sigma \frac{c_\sigma^2}{m_\sigma \omega_\sigma^2} \sum_{n=-\infty}^{\infty} \frac{\nu_n^2}{\nu_n^2 + \omega_\sigma^2} e^{i\nu_n \tau} = \frac{2}{\pi T} \int_0^\infty d\omega \frac{J(\omega)}{\omega} \sum_{n=-\infty}^{\infty} \frac{\nu_n^2}{\nu_n^2 + \omega^2} e^{i\nu_n \tau}$$

$$\tag{2.45}$$

, where the definition of the spectral density (2.39) has been used. The form of the kernel can be written in many forms (see [16]), here the zero temperature kernel $K(\tau)$ will be used which can be achieved after some minor manipulations of $k(\tau)$ [16]

$$K(\tau) = \frac{C}{2}\sum_{n=-\infty}^{\infty}\delta(\tau - nT) - k(\tau) = \frac{1}{T}\sum_{\sigma}\sum_{n=-\infty}^{\infty}\frac{c_\sigma^2}{m_\sigma(\nu_n^2 + \omega_\sigma^2)}e^{i\nu_n\tau} \quad (2.46)$$

where the summation over n is easily performed yielding in the principle interval $0 \leq \tau < T$ [16]

$$K(\tau) = \sum_{\sigma}\frac{c_\sigma^2}{2m_\sigma\omega_\sigma}\frac{\cosh\left(\omega_\sigma\left[\frac{T}{2}-\tau\right]\right)}{\sinh\left(\frac{\omega_\sigma T}{2}\right)} = \frac{1}{\pi}\int_0^\infty d\omega\, J(\omega)\frac{\cosh\left(\omega\left[\frac{T}{2}-\tau\right]\right)}{\sinh\left(\frac{\omega T}{2}\right)} \quad (2.47)$$

The resulting influence action with paths $q(\tau)$ extended outside the range of $0 \leq \tau < T$ using $q(\tau + nT) = q(\tau), \forall n \in \mathbb{N}$ is

$$S_{infl.}^E = \int_0^T d\tau \int_0^\tau d\tau'\, K(\tau - \tau')q(\tau)q(\tau')$$

$$K(\tau) \cong \sum_{\sigma}\frac{c_\sigma^2}{2m_\sigma\omega_\sigma}e^{-\omega_\sigma\tau} = \frac{1}{\pi}\int_0^\infty d\omega\, J(\omega)\, e^{-\omega\tau} \quad (2.48)$$

Since quantum dissipative effects due to the bath qualitatively depend on the large τ or low frequency behaviour, the fraction $\frac{\cosh\left(\omega\left[\frac{T}{2}-\tau\right]\right)}{\sinh\left(\frac{\omega T}{2}\right)}$ can be well approximated for small but fixed ω, large T and τ not to close to T,($\frac{T}{2}-\tau = \mathcal{O}(1)$), by $e^{-\omega\tau}$ as can be seen from the following calculation

$$\tau = \alpha\frac{T}{2}, 0 < \alpha < 1$$

$$\omega \neq 0: \lim_{T\to\infty}\frac{\cosh\left(\omega\left[\frac{T}{2}-\tau\right]\right)}{\sinh\left(\frac{\omega T}{2}\right)} = \lim_{T\to\infty}\frac{e^{\frac{\omega T}{2}(1-\alpha)} + e^{-\frac{\omega T}{2}(1-\alpha)}}{e^{\frac{\omega T}{2}} - e^{-\frac{\omega T}{2}}} = e^{-\omega\tau} \quad (2.49)$$

In this thesis the kernel $K(\tau)$ will be used to determine the position dependent tunnelling of the anharmonic bond(s). We will be able to show, that in the one anharmonic bond case there is ohmic dissipation ($\sim \tau^{-2}$) for the anharmonic bond located in the bulk of the chain, whereas there will be a transition from ohmic to super-ohmic ($\sim \tau^{-4}$) dissipation for the case of the anharmonic bond located at the border of the chain depending on a time scale defined by the position of the bond.

2.2.4 Instantons

The path integral formalism allows one to investigate quantum tunnelling by determining the instanton solutions [16], i.e. the solutions of the classical equation of motion for a double well

2.2. PHYSICAL PRELIMINARIES

potential **without** coupling to an environment, in imaginary time. The influence of the bath on the calculated instanton paths introduces the dissipative effect on the tunnelling. Getting the classical equations of motion from the Euclidean action S_0^E

$$S_0^E[q(\tau), \dot{q}(\tau)] = \int_0^T d\tau' L_0^E[q(\tau'), \dot{q}(\tau')]$$
$$L_0^E[q(\tau'), \dot{q}(\tau')] = \frac{m}{2}\dot{q}(\tau') + V_0(q(\tau')) \quad (2.50)$$

yields

$$\ddot{q}(\tau') - \frac{1}{m}\frac{\partial V_0(q(\tau'))}{\partial q(\tau')} = 0$$

Multiplying both sides by $\dot{q}(\tau')$ and integrating [12], we get

$$\int_0^T d\tau' \, \ddot{q}(\tau')\dot{q}(\tau') = \frac{1}{m}\int_0^T d\tau' \, \frac{\partial V_0(q(\tau'))}{\partial q(\tau')}\dot{q}(\tau') \quad (2.51)$$

, which is equivalent to

$$\frac{1}{2}\dot{q}^2 = \frac{V_0(q)}{m}$$
$$\Leftrightarrow \frac{dq(\tau')}{d\tau'} = \pm\sqrt{\frac{2V_0(q(\tau'))}{m}}$$
$$\Leftrightarrow \tau = \pm \int_{q(0)}^{q(\tau)} dq \sqrt{\frac{m}{2V_0(q)}} \quad (2.52)$$

Plugging in the symmetric double well potential $V_0(q) = \frac{C}{2}\left(q - \frac{q_0}{2}\right)^2\left(q + \frac{q_0}{2}\right)^2$ we get

$$\tau = \pm\sqrt{\frac{m}{C}} \int_{q(0)}^{q(\tau)} \frac{dq}{\left(q - \frac{q_0}{2}\right)\left(q + \frac{q_0}{2}\right)}$$
$$\tau = \pm\frac{2}{q_0}\sqrt{\frac{m}{C}} \text{Artanh}\left(\frac{2q(\tau')}{q_0}\right)\Big|_{\tau'=0}^{\tau'=\tau} \quad (2.53)$$

Choosing $q(0) = 0$ and inverting the Artanh, we get the instanton solution

$$q(\tau) = \pm\frac{q_0}{2}\tanh\left(\frac{q_0\tau}{2}\sqrt{\frac{C}{m}}\right) \quad (2.54)$$

By using the fact, that the upper phonon band edge ω_0 is roughly $\omega_0 \sim \sqrt{\frac{C}{m}}$ and assuming that $V_0''\left(\pm\frac{q_0}{2}\right) \approx C$, we can set the kink-width, the time an instanton needs to flip from one state to the other, $\tau_{kink} = \sqrt{\frac{m}{V_0''(\pm\frac{q_0}{2})}}$ to $\tau_{kink} \approx \omega_0^{-1}$. We are now able to write down the final form for the instanton solution for a symmetric double well potential

$$q(\tau) = \pm\frac{q_0}{2}\tanh\left(\frac{q_0}{2}\frac{\tau}{\tau_{kink}}\right) \cong \pm\frac{q_0}{2}\tanh\left(\frac{q_0}{2}\omega_0\tau\right) \qquad (2.55)$$

2.3 The Caldeira-Leggett Model

This section explains the Caldeira-Leggett model, which is used throughout this thesis. At first one has to know, that the Caldeira-Leggett Hamiltonian or Euclidean Lagrangian is originally of phenomenological nature. It is a model to describe a tunnelling quantum system at $T = 0$, linearly coupled to an environment. This model allows to observe quantum tunnelling on a macroscopic scale. The most promising candidate to see the effect of quantum tunnelling as a dominant factor in the transition is a SQUID (superconducting quantum interference device). The magnetic flux is the macroscopic variable in this scenario. WKB[1] approximations ignoring quantum dissipation do **not** show the full physical behaviour at low temperature compared to experiments. At large enough temperatures, the effect of thermal fluctuations affect the SQUID. Those corrections have been investigated by Kramers [31] and Kurkijärvi [32]. But these thermal fluctuations do **not** explain the experimentally observed deviations for low temperatures [24, 25]. This problem can be resolved by including quantum dissipative effects, which lead to a multiplicative factor in the tunnelling probability.

In 1981 Caldeira and Leggett proposed this simple model Euclidean Lagrangian [22] to describe the effect of quantum dissipation. The model Euclidean Lagrangian has the following form

$$L^E = \frac{m}{2}\dot{q}^2 + V(q) + \frac{1}{2}\sum_\sigma m_\sigma\left(\dot{Q}_\sigma^2 + \omega_\sigma^2 Q_\sigma^2\right) + q\sum_\sigma c_\sigma Q_\sigma \qquad (2.56)$$

where $\{Q_\sigma\}, \{\dot{Q}_\sigma\}$ denote coordinates and velocities of the harmonic environment and $q, \dot{q}$ the coordinate and velocity of the system. The parameters m_σ, ω_σ are at the masses and frequencies of the harmonic coordinates and c_σ are the coupling constants. With this Euclidean Lagrangian a propagator can be defined

$$G_E(q_f, \{Q_\sigma^{(f)}\}, T; q_i, \{Q_\sigma^{(i)}\}, 0) = \int_{q_i}^{q_f} \mathcal{D}q(\tau) \int_{\{Q_\sigma^{(i)}\}}^{\{Q_\sigma^{(f)}\}} \mathcal{D}\{Q_\sigma(t)\} \prod_\sigma e^{-\frac{1}{\hbar}\int_0^\tau d\tau' L^E} \qquad (2.57)$$

---
[1]Wentzel-Kramers-Brillouin

2.3. THE CALDEIRA-LEGGETT MODEL

where the harmonic degrees of freedom can be eliminated considering periodic paths $\{Q_\sigma^{(i)}\} = \{Q_\sigma^{(f)}\}$. Hence the propagator reads

$$G_E(q_f,\tau;q_i,0) = \int_{q_i}^{q_f} \mathcal{D}q(\tau)\, e^{-\frac{1}{\hbar}\int_0^\tau d\tau' \left(\frac{m}{2}\dot{q}^2 + V(q)\right)} e^{\frac{1}{\hbar}\int_{-\infty}^{\infty} d\tau' \int_0^{\tau'} d\tau''\, \alpha(\tau'-\tau'')q(\tau')q(\tau'') + \text{const.}} \quad (2.58)$$

where $q(\tau')$ has been periodically continued outside the region $0 < \tau' < \tau$ by the prescription $q(\tau'+\tau) \equiv q(\tau')$, which does not affect the tunnelling. The quantity $\alpha(\tau'-\tau'')$ is defined as follows

$$\alpha(\tau'-\tau'') = \sum_\sigma \frac{c_\sigma^2}{4m_\sigma\omega_\sigma} e^{-\omega_\sigma|\tau'-\tau''|} = \frac{1}{2\pi}\int_0^\infty d\omega\, J(\omega)e^{-\omega|\tau'-\tau''|} \quad (2.59)$$

where $J(\omega)$ is the spectral density defined in the subsection "Spectral Density". The constant is a term not contributing to the tunnelling and can be included into the potential $V(q)$ acting as a renormalization.

Now the main aspect of this model is to relate the quantity $\alpha(\tau'-\tau'')$ with the phenomenological friction coefficient η. Caldeira and Leggett note that since the characteristic times needed for the tunnelling are of order ω_{cl}^{-1}, or longer, hence $\alpha(\tau'-\tau'')$ is only needed for times of this order or expressed in terms of the spectral density $J(\omega)$ for frequencies $\omega \leq \omega_{cl}$. Now if the classical motion is to be determined by a well-defined friction coefficient, the following relation must hold

$$J(\omega \leq \omega_c) = \eta\omega \quad (2.60)$$

ω_c denotes a critical frequency, where the spectral density deviates appreciably from its low-frequency form, which is considered [22].

Using this restriction for the spectral density, the relation between quantity $\alpha(\tau'-\tau'')$ and the phenomenological friction coefficient is valid in lowest order of ω_{cl}/ω_c and reads

$$\int_{-\infty}^{\infty} d\tau' \int_0^{\tau'} d\tau''\, \alpha(\tau'-\tau'')q(\tau')q(\tau'') + \text{const.} = \frac{\eta}{4\pi}\int_{-\infty}^{\infty} d\tau' \int_0^{\tau'} d\tau''\, \frac{q(\tau')-q(\tau'')^2}{(\tau'-\tau'')} \quad (2.61)$$

A weakness of this model is, that it is purely phenomenological. The spectral density $J(\omega)$ is given a power law form ($\sim \omega^s$, $s > 0$) for low enough frequencies ($\omega \leq \omega_c$), but there is no microscopic model used to evaluate the spectral density analytically.

That is one of the main aspects of this thesis, to propose a model system/Hamiltonian, which can analytically be put into the Caldeira-Leggett form. This allows us, to microscopically derive the spectral density and hence the phenomenological input is not needed. The Caldeira-Leggett model has been researched quite intensively, but only for different exponents s of ω in the spectral density. The reason for this is the phase transition occurring at the critical value of $s = 1$. For $s < 1$ there is the localisation phenomenon. That means quantum tunnelling is fully suppressed, due to the interaction with the environment. For $s > 1$ quantum tunnelling is never suppressed. The effect of the environment is to damp the oscillation of the system. At the critical value $s = 1$ there is a phase transition from localisation to the damped oscillative behaviour, depending on the bonding constants.

2.4 Summary

In this chapter we have introduced mathematical and physical preliminaries, that are needed to fully understand the topics discussed in this thesis. The presentation of these definitions and techniques as a separate chapter will allow a more fluent reading of the main part. In this chapter only the basic techniques are presented, the derivation of the two anharmonic bond probability with the Feynman-Vernon path integral formalism, is shown here for one anharmonic bond in rough sketches. The full derivation for the more general case of two anharmonic bonds is presented in detail in the Appendix of the thesis.
In section "The Caldeira-Leggett Model" the model used in this thesis is presented. The model is motivated and its validity and possible applications are shown.

Chapter 3

The 1-defect Model

Lets consider a simple model of a 1-dimensional open chain of N-particles described by a classical Hamiltonian

$$H = \sum_{n=1}^{N} \frac{p_n^2}{2m_n} + V(x_1, ..., x_N)$$

$$V(x_1, ..., x_N) = \frac{C}{2} \sum_{\substack{n=1 \\ \neq M_i}}^{N-1} (x_{n+1} - x_n - a_n)^2 + \sum_{i=1}^{r} V_0(x_{M_i+1} - x_{M_i}) \quad (3.1)$$

where r are the number of anharmonic bonds, x_n is the position of the n-th particle, p_n the momentum of the n-th particle, C is the elastic constant of the harmonic nearest neighbour interaction, a_n are the equilibrium lengths of the harmonic bonds. The anharmonic potentials considered in this thesis are double well potentials with symmetric wells of the following form

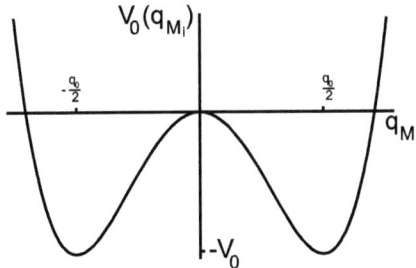

Figure 3.1: A symmetric double-well potential, for a one dimensional system. $q_{M_i} = x_{M_i+1} - x_{M_i}$, and with the local minima (ground states) $-\frac{q_0}{2}$ and $\frac{q_0}{2}$. The barrier height is labelled V_0

Here two methods of separating the harmonic from the anharmonic degrees of freedom are presented in sections "First Method" and "Second Method". The goal of the separation is to bring the Hamiltonian Eq. (3.1) into the form of the Caldeira-Leggett Hamiltonian or Euclidean Lagrangian, respectively

$$L^E = L_0^E + L_1^E, \quad L_1^E = L_{harm}^E + L_{infl}^E$$
$$L_0^E(q_M, \dot{q}_M) = \frac{M}{2}\dot{q}_M^2 + V_0(q_M)$$
$$L_1^E(q_M, \{Q_\sigma\}; \dot{q}_M, \{\dot{Q}_\sigma\}) = \frac{1}{2}\sum_{\sigma=1}^{N} m_\sigma \left[\dot{Q}_\sigma^2 + \omega_\sigma^2\left(Q_\sigma - \frac{c_\sigma}{m_\sigma \omega_\sigma^2}q_M\right)^2\right] \quad (3.2)$$

where the $\{Q_\sigma\}$ are the harmonic bath modes, that can be eliminated (the exact way to do this can be seen in the section "Physical Preliminaries"). The goal is to investigate the Euclidean propagator $G^E(\pm\frac{q_0}{2}, T; \mp\frac{q_0}{2}, 0)$, where the harmonic degrees of freedom have been eliminated, leading to an influence term in the Euclidean action

$$G_E\left(\pm\frac{q_0}{2}, T; \mp\frac{q_0}{2}, 0\right) = \int_{q_M(0)=\mp\frac{q_0}{2}}^{q_M(T)=\pm\frac{q_0}{2}} \mathcal{D}q_M(\tau)\, e^{-\frac{1}{\hbar}\left(S_0^E[q_M(\tau)] + S_{infl}^E[q_M(\tau)]\right)} \quad (3.3)$$

where the Euclidean action in the exponent is defined as

$$S_0^E[q_M(\tau)] = \int_0^T d\tau \left(\frac{m}{2}\dot{q}_M^2(\tau) + V_0(q_M(\tau))\right)$$
$$S_{infl.}^E[q_M(\tau)] = -\int_0^T d\tau \int_0^\tau d\tau'\, K(\tau - \tau')q_M(\tau)q_M(\tau') \quad (3.4)$$

This propagator describes the tunnelling of one classical minimum to the other during the observation time T. The kernel $K(\tau - \tau')$ shows the influence of the harmonic bath and will be derived in detail in section "Quantum Tunnelling". The main aspect being discussed here is the different behaviour of the kernel due to the position of the anharmonic bond. The kernel appears due to the elimination of the harmonic bath and leads to damping in the tunnelling behaviour of the anharmonic bond. Two cases are presented. One considers a macroscopically large chain, where the anharmonic bond is located at one of the borders of the chain with distance $\sim M$. Because of the translational invariance of the system, tunnelling of the anharmonic bond requires only a movement of a finite mass. In this case tunnelling is of super-ohmic dissipative nature and never suppressed. Whereas in the second case the anharmonic bond is located in the bulk of the macroscopically large chain. Tunnelling of the anharmonic bond requires now a movement of an infinite mass. For this case tunnelling is of ohmic dissipative nature and is fully suppressed if the coupling constant between the anharmonic bond and the harmonic bath exceeds a critical value. The details of what will happen are shown in section "Quantum Tunnelling".

In section "Two Defects" the separation of the harmonic and anharmonic degrees of freedom for two anharmonic bonds in one dimension is shown. The investigation of the kernel in the influence action is done to show the effect of interaction between the anharmonic bonds. Since only the effect of interaction between the two anharmonic bonds is of interest, both bonds are considered in the bulk with a finite distance D between each other. This scenario corresponds to an indirect interaction of both bonds via a finite harmonic bath between them and is discussed in detail in chapter "Two Defects".

The model Eq. (3.1) has been chosen so that an analytical discussion is possible without applying too many restrictions. Lets start with the case of one anharmonic bond ($r = 1$). First the two methods for separating the anharmonic from the harmonic degrees of freedom are presented.

3.1 First Method

The Hamiltonian Eq. (3.1) for one anharmonic bond ($r = 1$) in one dimension ($d = 1$) reads

$$H = \sum_{n=1}^{N} \frac{p_n^2}{2m_n} + V(x_1, ..., x_N)$$

$$V(x_1, ..., x_N) = \frac{C}{2} \sum_{\substack{n=1 \\ (\neq M)}}^{N-1} (x_{n+1} - x_n - a_n)^2 + V_0(x_{M+1} - x_M) \quad (3.5)$$

The first method is to introduce centre of mass of the *total* chain and relative coordinates. This approach is only applicable to 1d systems. The advantage of this approach is the way in which the harmonic and anharmonic degrees of freedom can be separated analytically and the resulting Hamiltonian be put in the form of the Euclidean Lagrangian of Caldeira and Leggett Eq. (3.2). Let

$$X_c = \frac{1}{M_c} \sum_{n=1}^{N} m_n x_n$$

$$M_c = \sum_{n=1}^{N} m_n \quad (3.6)$$

be the centre of mass and total mass for all particles and

$$q_i = \begin{cases} x_{i+1} - x_i - a_i & , \quad i = 1, ..., N-1 \\ X_C & , \quad n = 0 \end{cases}$$

$$a_M = 0 \quad (3.7)$$

be the relative coordinates, respectively. Now the canonical conjugate momenta have to be found. One can define a generating function (2.19)

$$R_2(x_1, ..., x_N; \pi_0, ..., \pi_{N-1}) = \sum_{k=0}^{N-1} \pi_k f_k(x_1, ..., x_N)$$

$$f_k(x_1, ..., x_N) = q_k(x_1, ..., x_N) + a_k \quad , \quad a_0 = 0 \quad (3.8)$$

This generating function (see section "Physical Preliminaries" (2.19)) defines the conjugate momenta as

$$p_n = \frac{\partial R_2}{\partial x_n} = \sum_{k=0}^{N-1} \pi_k \frac{\partial f_k(x_1, ..., x_N)}{\partial x_n} =: \sum_{k=0}^{N-1} M_{kn} \pi_k$$

$$M_{kn} = \frac{m_n}{M_c} \delta_{0,k} + (\delta_{k+1,n} - \delta_{k,n})(1 - \delta_{0,k}) \quad (3.9)$$

3.1. FIRST METHOD

where π_k are the canonical conjugate momenta of q_k. The matrix elements M_{nk} read explicitly

$$
\begin{array}{c}
\quad 1 \quad\; 2 \qquad\qquad\text{N-1}\quad\text{N}\\
\begin{array}{c}0\\1\\ \\ \\ \text{N-2}\\\text{N-1}\end{array}
\begin{pmatrix}
\frac{m_1}{M_c} & \frac{m_2}{M_c} & \cdots & \cdots & \cdots & \frac{m_N}{M_c}\\
-1 & 1 & 0 & \cdots & \cdots & 0\\
0 & -1 & 1 & \ddots & & \vdots\\
\vdots & \ddots & \ddots & \ddots & \ddots & \vdots\\
\vdots & & \ddots & -1 & 1 & 0\\
0 & \cdots & \cdots & 0 & -1 & 1
\end{pmatrix} = (M_{kn})
\end{array} \qquad (3.10)
$$

Applying the transformations for p_n of Eq. (3.9) to Eq. (3.5) yields

$$\sum_{n=1}^{N}\frac{p_n^2}{2m_n} = \frac{1}{2}\sum_{k,k'=0}^{N-1}\pi_k\pi_{k'}\sum_{n=1}^{N}\frac{1}{m_n}M_{kn}M_{k'n} = \frac{1}{2}\sum_{k,k'=0}^{N-1}\pi_k\pi_{k'}\underbrace{\sum_{n=1}^{N}\frac{1}{m_n}M_{kn}M_{nk'}^T}_{\widetilde{\mathbf{M}}} \qquad (3.11)$$

where $\widetilde{\mathbf{M}}$ is defined as

$$
\begin{array}{c}
\quad 0 \qquad\; 1 \qquad\qquad\qquad\qquad \text{N-2}\qquad\text{N-1}\\
\begin{array}{c}0\\1\\0\\ \vdots\\ \vdots\\ \text{N-2}\\ \text{N-1}\end{array}
\begin{pmatrix}
\frac{1}{M_c} & 0 & 0 & 0 & \cdots & \cdots & 0\\
0 & \frac{1}{m_1}+\frac{1}{m_2} & -\frac{1}{m_2} & 0 & \cdots & \cdots & 0\\
0 & -\frac{1}{m_2} & \frac{1}{m_2}+\frac{1}{m_3} & -\frac{1}{m_3} & 0 & \cdots & 0\\
\vdots & \ddots & \ddots & \ddots & \ddots & \ddots & \vdots\\
\vdots & & \ddots & \ddots & \ddots & \ddots & 0\\
0 & \cdots & \cdots & 0 & -\frac{1}{m_{N-2}} & \frac{1}{m_{N-2}}+\frac{1}{m_{N-1}} & -\frac{1}{m_{N-1}}\\
0 & \cdots & \cdots & \cdots & 0 & -\frac{1}{m_{N-1}} & \frac{1}{m_{N-1}}+\frac{1}{m_N}
\end{pmatrix} = \widetilde{\mathbf{M}}
\end{array}
$$

The resulting Hamiltonian is

$$H = \frac{\pi_0^2}{2M_c} + \frac{1}{2}\sum_{k=1}^{N-1}\left(\frac{1}{m_k}+\frac{1}{m_{k+1}}\right)\pi_k^2 - \sum_{k=1}^{N-2}\frac{\pi_k\pi_{k+1}}{m_{k+1}} + \frac{C}{2}\sum_{\substack{k=1\\(\neq M)}}^{N-1}q_k^2 + V_0(q_M) \qquad (3.12)$$

The first term is the kinetic energy of the centre of mass and will be dropped from now on. The Hamiltonian has to be diagonalised. To achieve this a canonical transformation decoupling the anharmonic momentum π_M from the harmonic momenta $\pi_k, k \neq M$ is performed. The coupling of π_M with $\pi_{M\pm1}$ reads

$$\frac{1}{2}\left(\frac{1}{m_M}+\frac{1}{m_{M+1}}\right)\pi_M^2 - \left(\frac{\pi_{M-1}}{m_M}+\frac{\pi_{M+1}}{m_{M+1}}\right)\pi_M =$$
$$\frac{1}{2}\frac{m_M+m_{M+1}}{m_M m_{M+1}}\left[\pi_M - \frac{m_{M+1}\pi_{M-1}+m_M\pi_{M+1}}{m_M+m_{M+1}}\right]^2 - \frac{(m_{M+1}\pi_{M-1}+m_M\pi_{M+1})^2}{2m_M m_{M+1}(m_M+m_{M+1})} \quad (3.13)$$

From this, the transformation decoupling the anharmonic momentum π_M from the harmonic momenta, can easily be seen

$$\tilde{p}_M = \pi_M - \frac{1}{m_M+m_{M+1}}(m_{M+1}\pi_{M-1}+m_M\pi_{M+1}) \quad (3.14)$$
$$\tilde{p}_k = \pi_k \quad , k=1,...,M-1, M+1,...,N-1 \quad (3.15)$$

Now the canonical conjugate coordinates have to be found. One can define a generating function (2.19)

$$R_3(\tilde{q}_1,...,\tilde{q}_{N-1};\pi_1,...,\pi_{N-1}) = -\tilde{q}_M\left[\pi_M - \frac{m_{M+1}\pi_{M-1}+m_M\pi_{M+1}}{m_M+m_{M+1}}\right]$$
$$-\sum_{\substack{k=1\\(\neq M)}}^{N-1}\tilde{q}_k\pi_k \quad (3.16)$$

With the generating function (2.19) the conjugate coordinates can be evaluated in the following way

$$q_k = -\frac{\partial R_3}{\partial \pi_k} = \tilde{q}_k \quad , k=1,...,M-2,M,M+2,...,N-1 \quad (3.17)$$
$$q_{M-1} = -\frac{\partial R_3}{\partial \pi_{M-1}} = \tilde{q}_{M-1} - \frac{1}{2}\tilde{q}_M \quad (3.18)$$
$$q_{M+1} = -\frac{\partial R_3}{\partial \pi_{M+1}} = \tilde{q}_{M+1} - \frac{1}{2}\tilde{q}_M \quad (3.19)$$

Substituting these new coordinates and momenta in Eq. (3.12) (note that the c.o.m. has been dropped) leads to

$$H = H_{harm} + H_d + H_{int}$$
$$H_{harm} = \frac{1}{2}\sum_{\substack{k,j=1\\(\neq M)}}^{N-1} T_{kj}\tilde{p}_k\tilde{p}_j + \frac{C}{2}\sum_{\substack{k=1\\(\neq M)}}^{N-1}\tilde{q}_k^2$$
$$H_d = \frac{m_M+m_{M+1}}{2m_M m_{M+1}}\tilde{p}_M^2 + \frac{C}{2}\frac{m_M^2+m_{M+1}^2}{(m_M+m_{M+1})^2}\tilde{q}_M^2 + V_0(q_M)$$
$$H_{int} = -C\frac{m_{M+1}\tilde{q}_{M-1}+m_M\tilde{q}_{M+1}}{m_M+m_{M+1}}q_M \quad (3.20)$$

3.1. FIRST METHOD

The Hamiltonian has been split up into three parts. The first one describes the harmonic part, where the matrix elements T_{kj} depend on the masses m_k and hence cannot be analytically diagonalised. Considering the simplest case of equal masses $m_k = m$ solves this problem and allows an analytical diagonalisation. The diagonalisation and the structure of T_{kj} for equal masses, is discussed in Appendix A. The second part describes the coordinates q_M, momenta $\tilde{p}_M$ and interaction $V_0(q_M)$. The last term contains the coupling of the harmonic with the anharmonic coordinates.

After the diagonalisation the introduction of normal coordinates

$$Q_\sigma = \sum_{\substack{k=1 \\ (\neq M)}}^{N-1} \tilde{q}_k u_k^{(\sigma)}, \quad P_\sigma = \sum_{\substack{k=1 \\ (\neq M)}}^{N-1} \tilde{p}_k u_k^{(\sigma)}, \quad \sigma = 1, ..., N-2 \qquad (3.21)$$

is possible. The eigenmodes $u_k^{(\sigma)}$ are derived and their explicit expressions are given in Eq. (A.11) in Appendix A. Applying Eq. (3.21) to H_{harm} of Eq. (3.20) yields

$$H_{harm} = \frac{1}{2} \sum_{\sigma=1}^{N-2} \left(\lambda_\sigma P_\sigma^2 + C Q_\sigma^2 \right) \qquad (3.22)$$

where λ_σ are the mass weighted eigenvalues defined in Eq. (A.10) in Appendix A. Applying the transformation to normal modes Eq. (3.21) to H_{int} of Eq. (3.20) yields

$$H_{int} = -\tilde{q}_M \sum_{\sigma=1}^{N-2} c_\sigma Q_\sigma \quad , \quad c_\sigma = \frac{C}{2} \left(u_{M+1}^{(\sigma)} + u_{M-1}^{(\sigma)} \right) \qquad (3.23)$$

Or expressed differently to make it easier to show the equivalence between the different approaches "First Method" and "Second Method" of separating the harmonic from the anharmonic degrees of freedom:

$$H_{int} = -\tilde{q}_M C \sum_{\sigma=1}^{N-2} \mathcal{N}_{b_\sigma} Q_\sigma \sin(q_\sigma M) \qquad (3.24)$$

, where the normalisation constant $\mathcal{N}_{b_\sigma}$ is defined in Appendix A (A.23). The term containing only the anharmonic degrees of freedom has the form

$$H_d = \frac{\tilde{p}_M^2}{m} + V_0(\tilde{q}_M) + \frac{C}{4} \tilde{q}_M^2 \qquad (3.25)$$

Comparing the harmonic Hamiltonian Eq. (3.22) with a standard harmonic Hamiltonian

$$H_{harm}^{(stand.)} = \frac{1}{2} \sum_\sigma \left[\frac{P_\sigma^2}{m_\sigma} + m_\sigma \omega_\sigma^2 Q_\sigma^2 \right] \qquad (3.26)$$

yields the following equations with $\omega_\sigma > 0$

$$m_\sigma = \frac{1}{\lambda_\sigma}$$
$$m_\sigma \omega_\sigma^2 = \frac{\omega_\sigma^2}{\lambda_\sigma} = C$$
$$\Rightarrow \omega_\sigma = \sqrt{C\lambda_\sigma} = \sqrt{\frac{2C}{m}(1-\cos(q_\sigma))} = 2\sqrt{\frac{C}{m}}\sin\left(\frac{q_\sigma}{2}\right) = \omega_0 \sin\left(\frac{q_\sigma}{2}\right) \quad (3.27)$$

, where ω_0 is the frequency of the upper phonon band edge. The final step now is to Legendre transform the Hamiltonian, which leads to a Lagrangian. After performing a Wick rotation ($t = -i\tau$) to the Lagrangian, the Euclidean Lagrangian reads

$$\begin{aligned} L^E &= L_0^E + L_1^E, \quad L_1^E = L_{harm}^E + L_{int}^E \\ L_0^E &= \frac{m}{4}\dot{q}_M^2 + V_0(q_M) \\ L_1^E &= \frac{1}{2}\sum_{\sigma=1}^{N-2} m_\sigma \left[\dot{Q}_\sigma^2 + \omega_\sigma^2 \left(Q_\sigma - \frac{c_\sigma}{m_\sigma \omega_\sigma^2} q_M\right)^2\right] \end{aligned} \quad (3.28)$$

This Euclidean Lagrangian is of the exact same form as the Caldeira-Leggett Euclidean Lagrangian presented in Eq. (3.2) needed to discuss the tunnelling behaviour. The completeness of the eigenvectors $u_k^{(\sigma)}$ leads to the equality

$$\sum_{\sigma=1}^{N-2} \frac{c_\sigma^2}{m_\sigma \omega_\sigma^2} \stackrel{(3.23),(3.27)}{=} \frac{C}{4} \sum_{\sigma=1}^{N-2} \left(u_{M+1}^{(\sigma)} + u_{M-1}^{(\sigma)}\right)^2 = \frac{C}{2} \quad (3.29)$$

This completeness Eq. (3.29) makes it possible to include the counter term (third term in H_d Eq. (3.25)) in L_1. This counter term, the role of which has been discussed by Weiss [16], results from the canonical transformations, Eq. (3.18)-(3.19). This transformation eliminates the coupling between the harmonic and the anharmonic momenta, as desired, and generates a coupling between the normal (harmonic-)coordinates $\{Q_\sigma\}$ and the corresponding system coordinate q_M. The Lagrangian now has the desired form for eliminating the normal (harmonic-)coordinates $\{Q_\sigma\}$ by the use of the path integral formalism. This procedure will be shown in section "Quantum Tunnelling".

3.2 Second Method

The second method of separating the harmonic and anharmonic degrees of freedom has the advantage that it can be applied to higher dimensional systems. Starting from the same Hamiltonian as in the first method Eq. (3.5) the centre of mass X_d is chosen for the anharmonic bond only, *not* for the whole chain as in method one. The relative system coordinate q_M stays equivalent to the first approach:

$$X_d = \frac{m_M x_M + m_{M+1} x_{M+1}}{m_M + m_{M+1}}$$
$$q_M = x_{M+1} - x_M \tag{3.30}$$

The corresponding canonical momenta can be achieved using a generating function (2.19) (like in the first approach). The generating function reads

$$R_2(x_M, x_{M+1}; \pi_M, P_d) = \pi_M q_M + P_d X_d \tag{3.31}$$

This generating function defines the conjugate momenta as

$$p_n = \frac{\partial R_2}{\partial x_n}, \quad n = M, M+1 \tag{3.32}$$

The resulting two equations can be put into the following form

$$P_d = p_M + p_{M+1}$$
$$\pi_M = \frac{m_M}{m_M + m_{M+1}} p_{M+1} - \frac{m_{M+1}}{m_M + m_{M+1}} p_M \tag{3.33}$$

Substituting these transformations Eqs. (3.30), (3.33) into Eq. (3.5) yields

$$H = H_d + H_{harm} + H_{int}$$
$$H_d = \frac{\pi_M^2}{2\mu_M} + V_0(q_M) + \frac{C}{2} \frac{m_M^2 + m_{M+1}^2}{(m_M + m_{M+1})^2} q_M^2$$
$$H_{harm} = \sum_{\substack{n=1 \\ (n \neq M, M+1)}}^{N} \frac{p_n^2}{2m_n} + \frac{P_d^2}{2(m_M + m_{M+1})} + \frac{C}{2} \sum_{\substack{n=1 \\ (n \neq M, M \pm 1)}}^{N-1} (x_{n+1} - x_n - a_n)^2$$
$$+ \frac{C}{2}(X_d - x_{M-1} - a_{M-1})^2 + \frac{C}{2}(x_{M+2} - X_d - a_{M+1})^2$$
$$H_{int} = -C \left[\frac{m_{M+1}}{m_M + m_{M+1}} (X_d - x_{M-1} - a_{M-1}) \right.$$
$$\left. + \frac{m_M}{m_M + m_{M+1}} (x_{M+2} - X_d - a_{M+1}) \right] q_M \tag{3.34}$$

where $\mu_M = m_M m_{M+1}/(m_M + m_{M+1})$ is the reduced mass of the anharmonic bond. For the transformation of H_{harm} Eq. (3.34) to normal coordinates, new coordinates are introduced for convenience

$$x'_n = \begin{cases} x_n & , \quad n = 1, ..., M-1 \\ X_d & , \quad n = M \\ x_{n+1} & , \quad n = M+1, ..., N-1 \end{cases}$$

$$p'_n = \begin{cases} p_n & , \quad n = 1, ..., M-1 \\ P_d & , \quad n = M \\ p_{n+1} & , \quad n = M+1, ..., N-1 \end{cases}$$

$$a'_n = \begin{cases} a_n & , \quad n = 1, ..., M-1 \\ a_{n+1} & , \quad n = M, ..., N-2 \end{cases}$$

$$m'_n = \begin{cases} m_n & , \quad n = 1, ..., M-1 \\ m_M + m_{M+1} & , \quad n = M \\ m_{n+1} & , \quad n = M+1, ..., N-1 \end{cases} \quad (3.35)$$

The explicit diagonalisation of the Hamiltonian by introducing again equal masses $m_n = m$, for the same reason as before, is done in Appendix B. This leads to

$$m'_n = m \begin{cases} 1 & , \quad n \neq M \\ 2 & , \quad n = M \end{cases} \quad (3.36)$$

A transformation to mass weighted normal modes

$$q'_\sigma = \sum_{n=1}^{N-1} \tilde{u}'_n e_n^{(\sigma)}, \quad p'_\sigma = \sum_{n=1}^{N-1} \tilde{p}'_n e_n^{(\sigma)} \quad (3.37)$$

with the eigenmodes $e_n^{(\sigma)}$ and $\tilde{u}'_n, \tilde{p}'_n$ defined in Appendix B, Eq. (B.5), yields:

$$\begin{aligned} H &= H_{harm} + H_{int} + H_d \\ H_{harm} &= \frac{1}{2} \sum_{\sigma=0}^{N-2} \left(p'^2_\sigma + \tilde{\lambda}_\sigma q'^2_\sigma \right) \\ H_{int} &= -q_M \sum_{\sigma=1}^{N-2} \tilde{c}_\sigma q'_\sigma \quad \tilde{c}_\sigma = \frac{C}{2\sqrt{m}} \left(e_{M+1}^{(\sigma)} - e_{M-1}^{(\sigma)} \right) \\ H_d &= \frac{\pi_M^2}{m} + V_0(q_M) + \frac{C}{4} q_M^2 \end{aligned} \quad (3.38)$$

, where $\tilde{\lambda}_\sigma$ are the eigenvalues defined in Appendix B, Eq. (B.9). Since H_{harm} (Eq. (3.38)) is still translationally invariant (note that only the c.o.m. for the defect has been separated of the total chain) there is a zero frequency mode which is chosen for $\sigma = 0$. The eigenvalue of

3.2. SECOND METHOD

the zero mode $\tilde{x}_\sigma$ reads $\tilde{\lambda}_0 = \frac{2C}{m}(1 - \cos(\tilde{x}_0)) = 0$.

After showing the equivalence of the transcendental equations (for the explicit calculation see Appendix B Eqs. (B.13)-(B.16)), only the equivalence of the Hamiltonian of the first approach Eqs. (3.22)-(3.25) with the Hamiltonian of the second method Eqs. (3.38) remains to be shown. Since H_d already has the same form, as can be seen by comparing H_d of the first method (Eq. (3.25) with H_d of the second method Eq. (3.38) (and replacing coordinates (π_M, q_M) by $(\tilde{p}_M, \tilde{q}_M)$), only the harmonic and interaction part remains. Extracting the zero frequency mode from the harmonic Hamiltonian of Eq. (3.38) one gets a harmonic and a c.o.m. part

$$H_{com} + H_{harm} = \frac{p_0'^2}{2} + \frac{1}{2}\sum_{\sigma=1}^{N-2}\left(p_\sigma'^2 + \tilde{\lambda}_\sigma q_\sigma'^2\right)$$

The c.o.m. Hamiltonian is the kinetic energy of the *total* chain. Note that the mass weighted eigenvalues from the first method λ_σ can be transformed into the mass weighted eigenvalues of the second method $\tilde{\lambda}_\sigma$ by

$$\tilde{\lambda}_\sigma = C\lambda_\sigma \tag{3.39}$$

With the canonical transformations

$$p_\sigma' := \sqrt{\lambda_\sigma}P_\sigma, \quad q_\sigma' := \frac{1}{\sqrt{\lambda_\sigma}}Q_\sigma \qquad \sigma = 1,...,N-2 \tag{3.40}$$

the same harmonic Hamiltonian as in the first approach

$$H_{harm} = \frac{1}{2}\sum_{\sigma=1}^{N-2}\left(\lambda_\sigma P_\sigma^2 + CQ_\sigma^2\right) \tag{3.41}$$

is achieved. The c.o.m. momentum p_0' in the second approach (which is mass weighted) is of course nothing put the massless momentum of the first approach π_0 with the mass added (in a multiplicative way) separately (see Eq. (3.12)). The transformation reads

$$p_0' = \frac{\pi_0}{\sqrt{M_c}} \tag{3.42}$$

This shows the equivalence of the c.o.m. and harmonic Hamiltonian of both approaches. The interaction part is more tedious. Starting with the interaction part of Eq. (3.38) and using the transformation given in Eq. (3.40) one gets

$$H_{int} = -q_M \frac{C}{2\sqrt{m}}\sum_{\sigma=1}^{N-2}\frac{Q_\sigma}{\sqrt{\lambda_\sigma}}\left(e_{M+1}^{(\sigma)} - e_{M-1}^{(\sigma)}\right) \tag{3.43}$$

By using the derived eigenmodes from Eqs. (B.9), (B.10) one gets the following expression for H_{int}

$$H_{int} = -q_M \frac{C}{2\sqrt{m}} \sum_{\sigma=1}^{N-2} \tilde{\mathcal{N}}_{\tilde{b}_\sigma} \frac{Q_\sigma}{\sqrt{\lambda_\sigma}} \left(\frac{\cos\left(\tilde{x}_\sigma \left[M - \frac{1}{2}\right]\right) \cos\left(\tilde{x}_\sigma \left[N - M - \frac{3}{2}\right]\right)}{\cos\left(\tilde{x}_\sigma \left[N - M - \frac{1}{2}\right]\right)} - \cos\left(\tilde{x}_\sigma \left[M - \frac{3}{2}\right]\right) \right)$$

Now with the help of the transcendental equation Eq. (B.13) one can replace the first term in parenthesis by $2\left(2\cos(\tilde{x}_\sigma) - 1\right) \cos\left(\tilde{x}_\sigma \left[M - \frac{1}{2}\right]\right) - \cos\left(\tilde{x}_\sigma \left[M - \frac{3}{2}\right]\right)$. This allows us to rewrite the interaction part as

$$H_{int} = -q_M \frac{C}{\sqrt{m}} \sum_{\sigma=1}^{N-2} \tilde{\mathcal{N}}_{\tilde{b}_\sigma} \frac{Q_\sigma}{\sqrt{\lambda_\sigma}} \left(\left(2\cos(\tilde{x}_\sigma) - 1\right) \cos\left(\tilde{x}_\sigma \left[M - \frac{1}{2}\right]\right) - \cos\left(\tilde{x}_\sigma \left[M - \frac{3}{2}\right]\right) \right)$$

With the help of some basic trigonometric identities we get

$$H_{int} = 2q_M \frac{C}{\sqrt{m}} \sum_{\sigma=1}^{N-2} \tilde{\mathcal{N}}_{\tilde{b}_\sigma} \frac{Q_\sigma}{\sqrt{\lambda_\sigma}} \sin\left(\tilde{x}_\sigma M\right) \sin\left(\frac{\tilde{x}_\sigma}{2}\right) \tag{3.44}$$

using the equation for the eigenvalues Eqs. (3.39), (B.9) the interaction part has exactly the same form as in the first method Eq. (3.24).

$$H_{int} = -q_M C \sum_{\sigma=1}^{N-2} \tilde{\mathcal{N}}_{\tilde{b}_\sigma} Q_\sigma \sin\left(\tilde{x}_\sigma M\right) \tag{3.45}$$

To show the full equivalence one has to look closer at the variables. Both approaches yield the same Q_σ and of course $\tilde{q}_M = q_M$. The equivalence of the transcendental equations (proven in Appendix B Eq. (B.16)) also provide the same wave numbers $q_\sigma = \tilde{x}_\sigma$ for both approaches. The only equivalence left to prove is that of the normalisation constants. Since this is not only tedious, but also more to write down, the proof is given at the end of Appendix B. With this result the equivalence of both approaches has been shown.

3.3 Quantum Tunnelling

Since both approaches lead to the same Euclidean Lagrangian, the notation of the first approach is used from now on. This section is restricted to the zero temperature quantum tunnelling behaviour of the anharmonic bond embedded in the harmonic chain. The tunnelling behaviour is graphically shown in Figure (3.2)

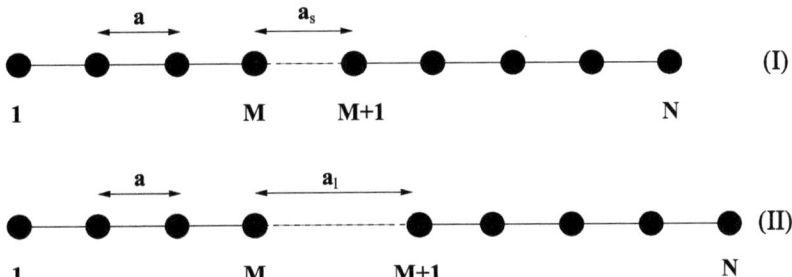

Figure 3.2: Two degenerate classical ground states of the open chain with N particles. The masses m_n are chosen to be equal. a is the equilibrium length of the harmonic bonds and a_s, a_l the two degenerate equilibrium lengths of the anharmonic bond.

The defect potential $V_0(q_M)$ of Eq. (3.28) is assumed to be a double-well potential with degenerate minima at $-\frac{q_0}{2}$, which corresponds to the equilibrium length $a_s > 0$ of the anharmonic bond and $\frac{q_0}{2}$, which corresponds to the other equilibrium length $a_l > a_s$ of the anharmonic bond. The interest lies in calculating the Euclidean propagator describing the tunnelling between the degenerated equilibrium ground states $-\frac{q_0}{2}$ and $\frac{q_0}{2}$. The propagator describing this tunnelling behaviour is defined as

$$G_E(q_f, T; q_i, 0) = \int_{q_M(0)=q_i}^{q_M(T)=q_f} \mathcal{D}q_M(\tau)\, e^{-\frac{S^E[q_M(\tau)]}{\hbar}} \qquad (3.46)$$

where the harmonic degrees of freedom have already been eliminated[1]. The Euclidean action

---

[1] The elimination procedure for the harmonic degrees of freedom has been shown explicitly in section "Physical Preliminaries"

$S^E[q_M(\tau)]$ for this model can be split up into two parts

$$S^E[q_M(\tau)] = S_0^E[q_M(\tau)] + S_{infl.}^E[q_M(\tau)]$$

$$S_0^E[q_M(\tau)] = \int_0^T d\tau \left(\frac{m}{2}\dot{q}_M^2(\tau) + V(q_M(\tau))\right)$$

$$S_{infl.}^E[q_M(\tau)] = -\int_0^T d\tau \int_0^\tau d\tau' K_M(\tau-\tau') q_M(\tau) q_M(\tau') \tag{3.47}$$

$$K_M(\tau) = \sum_{\sigma=1}^{N-2} \frac{c_\sigma^2}{2m_\sigma \omega_\sigma} e^{-\omega_\sigma \tau} \tag{3.48}$$

The kernel $K_M(\tau)$ will now be discussed to show the tunnelling behaviour for this case. The index M stands for the position dependence of the anharmonic bond. Inserting c_σ from Eq. (3.23) and m_σ, ω_σ from Eqs. (3.27) into the kernel (3.48) yields

$$K_M(\tau) = \frac{C\omega_0}{2} \sum_{\sigma=1}^{N-2} \mathcal{N}_{b_\sigma}^2 \sin\left(\frac{q_\sigma}{2}\right) \sin^2(q_\sigma M) \, e^{-\omega_0 \sin(\frac{q_\sigma}{2})\tau} \tag{3.49}$$

Replacing the normalisation constant $\mathcal{N}_{b_\sigma}$ (A.23) by its low-frequency behaviour $\sqrt{\frac{2}{N}}$ which is a valid assumption for the case of taking the thermodynamic limit ($N \to \infty$), yields the integral representation of the kernel

$$K_M(\tau) \cong \frac{C\omega_0}{2\pi} \int_0^\pi dq \, q \sin^2(qM) \, e^{-\frac{\omega_0 q}{2}\tau} \tag{3.50}$$

From the integrand of Eq. (3.50) two q-scales follow. They are

$$q_M = \frac{1}{M} \quad \text{and} \quad q_\tau = \frac{1}{\omega_0 \tau} \tag{3.51}$$

Equating $q_M = q_\tau$ defines a time scale

$$\tau_M = \frac{M}{\omega_0} \tag{3.52}$$

The kernel Eq. (3.50) can be evaluated by performing the q-integration, the result is

$$K_M(\tau) \cong \frac{8CM^2\omega_0 e^{-\frac{\pi}{2}\omega_0\tau}\left(16M^2\left[2e^{\frac{\pi}{2}\omega_0\tau} - 2 - \pi\omega_0\tau\right] - (\omega_0\tau)^2\left[6 - 6e^{\frac{\pi}{2}\omega_0\tau} + \pi\omega_0\tau\right]\right)}{\pi(16M^2\omega_0\tau + (\omega_0\tau)^3)^2}$$

This result can be approximated neglecting the exponentially decaying factors regarding the long time limit $\omega_0 \tau \gg 1$

$$K_M(\tau) \cong \frac{8CM^2\omega_0\left(32M^2 + 6(\omega_0\tau)^2\right)}{\pi(16M^2\omega_0\tau + (\omega_0\tau)^3)^2} \tag{3.53}$$

This memory kernel now has to be investigated regarding the above defined two time scales.

3.3. QUANTUM TUNNELLING

- $\tau \gg \tau_M$:

  Replacing M by $\omega_0 \tau_M$ and applying $\tau \gg \tau_M$, yields

  $$K_M(\tau \gg \tau_M) \cong \frac{48 C \omega_0 M^2}{\pi} \frac{1}{(\omega_0 \tau)^4} \qquad (3.54)$$

- $\tau \ll \tau_M$:

  In this case τ_M yields the largest contribution, applying this to Eq. (3.53)

  $$K_M(\tau \ll \tau_M) \cong \frac{C \omega_0}{\pi} \frac{1}{(\omega_0 \tau)^2} \qquad (3.55)$$

  which leads to ohmic behaviour.

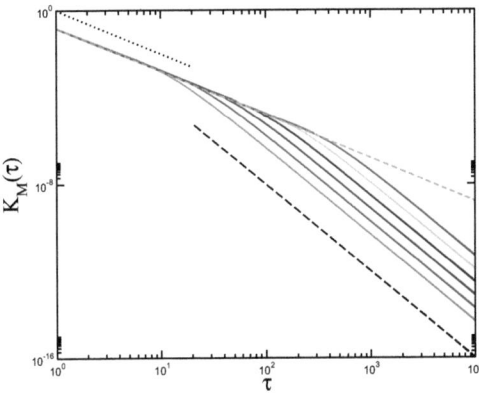

Figure 3.3: τ-dependence of $K_M(\tau)$ for different finite and infinite M (green: $M = 5$, red: $M = 10$, blue: $M = 20$, brown: $M = 40$, grey: $M = 80$, magenta: $M = 160$, dashed orange: $M = \infty$) on a log-log representation. The dotted and dashed line, corresponding to τ^{-2} and τ^{-4}, respectively included to see the transition.

Finally one can summarise the results as follows. If the observation time T of the propagator (3.46) is smaller than τ_M, Eq. (3.55) holds, which shows ohmic dissipation. This can be explained as follows. τ_M is the time a phonon emitted from the anharmonic bond during a change of length, needs to reach one of the chain ends. If the anharmonic bond is located in the bulk of a macroscopically large chain, it never *feels* the ends of the chain, because τ_M is infinitely large and thus τ is always smaller τ_M. That means the dissipation for an anharmonic bond located in the bulk of an infinitely large set of harmonic oscillators with linear coupling, is always ohmic.

Figure 3.3 shows the Kernel $K_M(\tau)$ in a log-log-plot. The transition from ohmic disipative behaviour $\sim \tau^{-2}$ for $\tau < \tau_M$ to superohmic dissipative behaviour $\sim \tau^{-4}$ is clearly visible in Figure 3.3 for different values of M. For the case of ohmic dissipation, which means observation times $T < \tau_M$, there exists a phase transition allowing a mapping of this problem on the one dimensional Ising model with long range interactions of the form $\frac{1}{\tau^2}$, as has been performed by [33] showing this phase transition.

There exists a critical coupling $C_{crit}(T)$ which separates the ordered phase $C > C_{crit}(T)$ from the disordered phase $C < C_{crit}(T)$. The parameter T represents the *length* of the Ising chain. The expression *feeling* (used above) can be interpreted in terms of the Ising chain as a correlation length $\xi(C)$. In the Ising model the correlation length is usually given as a function of the temperature [34] (section III 17.1), but since the temperature is zero here, the relevant variable is the coupling C. Since the one dimensional Ising model does not show a sharp transition for a finite length T, there is also no sharp transition from tunnelling $C < C_{crit}(T)$ to "long range order" at $C > C_{crit}(T)$, which corresponds to localisation. A sharp transition (phase transition) can be observed for $T = \infty$, only.

Chapter 4

The 2-defect Model

4.1 Two Defects

The case of an open linear chain of N particles with next neighbour interactions and two defects M_1, M_2 is considered. The Hamiltonian for this system reads as

$$H = \sum_{n=1}^{N} \frac{p_n^2}{2m_n} + \frac{C}{2} \sum_{\substack{n=1 \\ (\neq M_1, M_2)}}^{N-1} (x_{n+1} - x_n - a_n)^2 + \sum_{i=1}^{2} V_0(x_{M_i+1} - x_{M_i}) \qquad (4.1)$$

Separating the harmonic from the anharmonic degrees of freedom is done in the same way and for the same reasons as in the first method of the one anharmonic bond case (see Eq. (3.14)-(3.19)). Let

$$X_c = \frac{1}{M_c} \sum_{n=1}^{N} x_n m_n$$

$$M_c = \sum_{n=1}^{N} m_n \qquad (4.2)$$

be the centre of mass and the total mass for all particles and

$$y_n = x_{n+1} - x_n - a_n \quad , \quad n = 1, ..., N-1; n \neq M_1, M_2 \qquad (4.3)$$

be the relative coordinates, respectively. As in the one anharmonic bond case, one can define a generating function (since the procedure is **absolutely** equivalent (3.8), the calculation will not be shown here). For convenience the simplest case making an analytical diagonalisation possible $m_n = m$, is chosen. It yields

$$H = \frac{\pi_0^2}{2M_c} + \frac{1}{m} \left(\sum_{k=1}^{N-1} \pi_k^2 - \sum_{k=1}^{N-2} \pi_k \pi_{k+1} \right) + \frac{C}{2} \sum_{\substack{k=1 \\ (\neq M_1, M_2)}}^{N-1} y_k^2 + V_0(y_{M_1}) + V_0(y_{M_2}) \qquad (4.4)$$

This Hamiltonian is the starting point for the diagonalisation procedure. One assumption needed to perform the analytical diagonalisation in the way shown in Appendix C is to define the middle of the open chain M as either $M = \frac{N}{2}$ for an even chain-length and $M = \frac{N-1}{2}$ for an odd chain-length. Choosing the anharmonic bonds symmetrically located with respect to M, one can express the variables M_1, M_2, N in terms of N, D, where $D = M_2 - M_1$ is the difference between both bonds and $N = M_1 + M_2$ is the total chain-length. This approximation does not qualitatively change the tunnelling behaviour, since both anharmonic bonds are only investigated in the bulk. In the next subsection the relevant canonical transformations for $D \geq 2$ are presented. That covers all possible cases, since the case of $D = 1$ means that both anharmonic bonds are coupled directly and could be considered as one anharmonic bond with an additional degree of freedom. This case will not be considered, since the indirect interaction between both anharmonic bonds through the harmonic bath is of interest.

4.1.1 Pre-diagonalisation Transformations

Again canonical transformations for the diagonalisation procedure are applied (see explanation before Eqs. (3.14)-(3.19)). The decoupling of both anharmonic bonds from the harmonic degrees of freedom is done analogously to section "First Method". The generating function is achieved in the same way as before and reads

$$R_3(q_1, ..., q_{N-1}; \pi_1, ..., \pi_{N-1}) = - q_{M_i} \left[\pi_{M_i} - \frac{\pi_{M_i+1} + \pi_{M_i-1}}{2} \right]$$
$$- \sum_{\substack{k=1 \\ (\neq \{M_i\})}}^{N-1} q_k \pi_k \quad , i = 1, 2 \quad (4.5)$$

With this generating function the conjugate coordinates can be evaluated in the following way

$$q_k = -\frac{\partial R_3}{\partial \pi_k} = y_k \quad , k \neq M_i \pm 1 \quad (4.6)$$
$$q_{M_i \pm 1} = -\frac{\partial R_3}{\partial \pi_{M_i \pm 1}} = y_{M_i \pm 1} - \frac{1}{2} y_{M_i} \quad , i = 1, 2 \quad (4.7)$$

4.1. TWO DEFECTS

The transformations for $D = M_2 - M_1 \geq 2$ read

$$\begin{aligned}
\pi_k &= p_k \quad , \quad k=1,...,N-1 \quad ; k \neq M_1, M_2 \\
\pi_{M_1} &= p_{M_1} + \frac{1}{2}(p_{M_1+1} + p_{M_1-1}) \\
\pi_{M_2} &= p_{M_2} + \frac{1}{2}(p_{M_2+1} + p_{M_2-1}) \\
y_k &= q_k \quad , \quad k=1,...,N-1 \quad ; k \neq M_1 \pm 1, M_2 \pm 1 \\
y_{M_1 \pm 1} &= q_{M_1 \pm 1} - \frac{q_{M_1}}{2} \\
y_{M_2 \pm 1} &= q_{M_2 \pm 1} - \frac{q_{M_2}}{2}
\end{aligned} \quad (4.8)$$

Substituting Eq. (4.8) into Eq. (4.4) the Hamiltonian can be split up into four parts

$$\begin{aligned}
H &= H_{com} + H_{harm} + H_{int} + H_d \\
H_{com} &= \frac{\pi_0^2}{2M_c} \\
H_{harm} &= \frac{1}{m}\left[\sum_{\substack{k=1 \\ (\neq \{M_i\})}}^{N-1} p_k^2 - \sum_{\substack{k=1 \\ (\neq \{M_i\},\{M_i-1\})}}^{N-2} p_k p_{k+1} - \frac{1}{4}\sum_{i=1}^{2}(p_{M_i+1} + p_{M_i-1})^2\right] \\
&\quad + \frac{C}{2}\sum_{\substack{k=1 \\ (\neq \{M_i\})}}^{N-1} q_k^2 \\
H_{int} &= -\frac{C}{2}q_{M_1}\left(q_{M_1-1} + q_{M_1+1}\right) - \frac{C}{2}q_{M_2}\left(q_{M_2-1} + q_{M_2+1}\right) \\
H_d &= \frac{1}{m}\left(p_{M_1}^2 + p_{M_2}^2\right) + \frac{C}{4}\left(q_{M_1}^2 + q_{M_2}^2\right) + V_0(q_{M_1}) + V_0(q_{M_2})
\end{aligned} \quad (4.9)$$

The c.o.m. term will be dropped from now on. It just represents the translational invariance of the system and does not contribute to the influence action. The relevant term for the diagonalisation T_{harm} (kinetic part of the harmonic Hamiltonian) will now be treated. The explicit derivation is presented in detail in Appendix C.

4.1.2 Caldeira-Leggett form for $D \geq 2$

After the diagonalisation a transformation to normal modes as done in (3.21)

$$\begin{aligned}
p_k &= \sum_{\sigma=1}^{N-3} P_\sigma^{(\alpha)} u_k^{(\sigma),\alpha} \quad , \quad k \neq M_1, M_2 \\
q_k &= \sum_{\sigma=1}^{N-3} Q_\sigma^{(\alpha)} u_k^{(\sigma),\alpha} \quad , \quad k \neq M_1, M_2
\end{aligned} \quad (4.10)$$

where $\vec{u}^{(\sigma),\alpha}$ denote the eigenvectors achieved in the diagonalisation procedure. The additional index α can take two values "symmetric" and "antisymmetric" and is due to the diagonalisation procedure performed in Appendix C. That transformation applied to Eq. (4.9) yields (the c.o.m. term has been omitted)

$$\begin{aligned}
H_{harm} &= \frac{1}{2m} \sum_{\substack{k,l=1 \\ (\neq M_1,M_2)}}^{N-1} T_{kl} p_k p_l + \frac{C}{2} \sum_{\substack{k=1 \\ (\neq M_1,M_2)}}^{N-1} q_k^2 \\
&= \frac{1}{2} \sum_\alpha \sum_{\substack{k=1 \\ (\neq M_1,M_2)}}^{N-1} \sum_{\sigma,\sigma'=1}^{N-3} \lambda_\sigma^{(\alpha)} P_\sigma^{(\alpha)} P_{\sigma'}^{(\alpha)} u_k^{(\sigma),\alpha} u_k^{(\sigma'),\alpha} \\
&+ \frac{C}{2} \sum_\alpha \sum_{\substack{k=1 \\ (\neq M_1,M_2)}}^{N-1} \sum_{\sigma,\sigma'=1}^{N-3} Q_\sigma^{(\alpha)} Q_{\sigma'}^{(\alpha)} u_k^{(\sigma),\alpha} u_k^{(\sigma'),\alpha}
\end{aligned} \quad (4.11)$$

where T_{kl} denotes the matrix elements and $\lambda_\sigma^{(\alpha)}$ the eigenvalues used in Eq. (C.10) for the diagonalisation. Performing the summation over k simplifies the expression yielding

$$H_{harm} = \frac{1}{2} \sum_\alpha \sum_{\sigma=1}^{N-3} \left[\lambda_\sigma^{(\alpha)} \left(P_\sigma^{(\alpha)}\right)^2 + C \left(Q_\sigma^{(\alpha)}\right)^2 \right] \quad (4.12)$$

The interaction part reads

$$H_{int} = -\sum_\alpha \sum_{\sigma=1}^{N-3} Q_\sigma^{(\alpha)} \left[c_{1,\sigma}^\alpha q_{M_1} + c_{2,\sigma}^\alpha q_{M_2} \right], \quad c_{i,\sigma}^\alpha = \frac{C}{2}\left(u_{M_i+1}^{(\sigma),\alpha} + u_{M_i-1}^{(\sigma),\alpha} \right), \quad i=1,2 \quad (4.13)$$

and finally the defect part stays as it was

$$H_d = \frac{1}{m}\left(p_{M_1}^2 + p_{M_2}^2\right) + \frac{C}{4}\left(q_{M_1}^2 + q_{M_2}^2\right) + V_0(q_{M_1}) + V_0(q_{M_2}) \quad (4.14)$$

A Legendre transformation and some basic mathematical manipulation lead to the desired Euclidean Lagrangian in the form of Caldeira-Leggett

$$\begin{aligned}
L^E &= L_0^E + L_1^E \\
L_0^E &= \frac{m}{4}\left[\dot{q}_{M_1}^2 + \dot{q}_{M_2}^2\right] + V_0(q_{M_1}) + V_0(q_{M_2}) \\
L_1^E &= \frac{1}{2} \sum_\alpha \sum_{\sigma=1}^{N-3} m_\sigma^\alpha \left[\left(\dot{Q}_\sigma^{(\alpha)}\right)^2 + \left(\omega_\sigma^{(\alpha)}\right)^2 \left(Q_\sigma^{(\alpha)}\right)^2 \right] - \sum_\alpha \sum_{\sigma=1}^{N-3} Q_\sigma^{(\alpha)} \left(c_{1,\sigma}^\alpha q_{M_1} + c_{2,\sigma}^\alpha q_{M_2}\right) \\
&+ \sum_\alpha \sum_{\sigma=1}^{N-3} \frac{\left(c_{1,\sigma}^\alpha q_{M_1}\right)^2 + \left(c_{2,\sigma}^\alpha q_{M_2}\right)^2}{2 m_\sigma^\alpha \left(\omega_\sigma^{(\alpha)}\right)^2}
\end{aligned} \quad (4.15)$$

4.1. TWO DEFECTS

where

$$m_\sigma^\alpha = \frac{1}{\lambda_\sigma^{(\alpha)}} \quad , \quad \omega_\sigma^{(\alpha)} = \sqrt{C\lambda_\sigma^{(\alpha)}} = \omega_0 \sin\left(\frac{q_\sigma^{(\alpha)}}{2}\right) \quad , \quad \omega_0 = 2\sqrt{\frac{C}{m}} \qquad (4.16)$$

are achieved in the same way and for $\omega_\sigma^{(\alpha)} > 0$ as in Eq. (3.27). The coupling coefficients read with the eigenvectors from Eq. (C.9)

$$\begin{aligned}
c_{1,\sigma}^a &= -c_{2,\sigma}^a = C\mathcal{N}_{\sigma,a} \sin\left(q_\sigma^{(a)} \frac{N-D}{2}\right) \\
c_{1,\sigma}^s &= c_{2,\sigma}^s = C\mathcal{N}_{\sigma,s} \sin\left(q_\sigma^{(s)} \frac{N-D}{2}\right)
\end{aligned} \qquad (4.17)$$

and are necessary for the calculation of the zero temperature kernel. The normalisation constants have been calculated in Appendix C, Eqs. (C.27), (C.28).

4.1.3 The Kernel for two anharmonic Bonds

To discuss the tunnelling behaviour for two anharmonic bonds, the propagator in form of an Euclidean Green's function as in the one anharmonic bond case is needed. The propagator reads

$$G_E(q_{M_1}^f, q_{M_2}^f, T; q_{M_1}^i, q_{M_2}^i, 0) = \int_{q_{M_1}(0)=q_{M_1}^i}^{q_{M_1}(T)=q_{M_1}^f} \mathcal{D}q_{M_1}(\tau) \int_{q_{M_2}(0)=q_{M_2}^i}^{q_{M_2}(T)=q_{M_2}^f} \mathcal{D}q_{M_2}(\tau)\, e^{-\frac{S^E[q_{M_1}(\tau),q_{M_2}(\tau)]}{\hbar}} \qquad (4.18)$$

Splitting the Euclidean action into a local and an influence part as done in Eq. (3.4), where the local part yields the instanton solutions and the influence part the kernel, we are able to write down the influence action for the two defect case (derivation analogous to the one anharmonic bond case see Eq. (3.47)) as

$$S_{infl.}^E[q_{M_1}, q_{M_2}] = -\sum_{i,j=1}^{2} \int_0^T d\tau \int_0^\tau d\tau'\, K_D^{ij}(\tau-\tau') q_{M_i}(\tau) q_{M_j}(\tau') \qquad (4.19)$$

where the indices i, j show the effect of two instead of one anharmonic bond, as discussed before. The derivation of the zero temperature kernel for the two anharmonic bond case can be achieved from the Euclidean Lagrangian in exactly the same way as before (see Section: "Physical Preliminaries"). The calculation will not be presented here. The result is:

$$K_D^{ij}(\tau) = \sum_\alpha K_D^{(\alpha),ij}(\tau) \cong \sum_\alpha \sum_{\sigma=1}^{N-3} \frac{c_{i,\sigma}^{(\alpha)} c_{j,\sigma}^{(\alpha)}}{2m_\sigma^{(\alpha)} \omega_\sigma^{(\alpha)}} e^{-\omega_\sigma^{(\alpha)}\tau} \qquad (4.20)$$

The coupling of symmetric with antisysmmetric eigenvectors is of course zero by definition, hence $c_{i,\sigma}^{(s)} c_{j,\sigma}^{(a)} = 0$. That is the reason only $c_{i,\sigma}^{(\alpha)} c_{j,\sigma}^{(\alpha)}$ is considered in Eq. (4.20). The calculation of the kernel is only necessary for one symmetric and one antisymmetric case, because the kernel obeys, as can be seen from Eqs. (4.17) the following relations

$$\begin{aligned}K_D^{(s),11}(\tau) &= K_D^{(s),12}(\tau) = K_D^{(s),21}(\tau) = K_D^{(s),22}(\tau)\\ K_D^{(a),11}(\tau) &= -K_D^{(a),12}(\tau) = -K_D^{(a),21}(\tau) = K_D^{(a),22}(\tau)\end{aligned} \quad (4.21)$$

As discussed in Appendix C, the thermodynamic limit $N \to \infty$ makes an even chainlength indistinguishable from an odd chainlength. This is the reason only even N are considered in the calculations and derivations, respectively. Starting with the symmetric case using Eqs. (4.16), (4.17), the kernel reads as

$$\begin{aligned}K_D^{(s),ij}(\tau) &= \frac{C\omega_0}{2} \sum_\sigma \mathcal{N}_{\sigma,s}^2 \sin^2\left(q_\sigma^{(s)} \frac{N-D}{2}\right) \sin\left(\frac{q_\sigma^{(s)}}{2}\right) e^{-\omega_0 \sin\left(\frac{q_\sigma^{(s)}}{2}\right)\tau}\\ &= \frac{C\omega_0}{2} \sum_\sigma \mathcal{N}_{\sigma,s}^2 \sin\left(\frac{q_\sigma^{(s)}}{2}\right) e^{-\omega_0 \sin\left(\frac{q_\sigma^{(s)}}{2}\right)\tau}\\ &\quad \cdot \left[\sin\left(\frac{q_\sigma^{(s)}N}{2}\right)\cos\left(\frac{q_\sigma^{(s)}D}{2}\right) - \cos\left(\frac{q_\sigma^{(s)}N}{2}\right)\sin\left(\frac{q_\sigma^{(s)}D}{2}\right)\right]^2\end{aligned}$$

Now $\sin\left(\frac{q_\sigma^{(a)}N}{2}\right), \cos\left(\frac{q_\sigma^{(a)}N}{2}\right)$ can be replaced with the results obtained from the transcendental equations (C.30), and since the main contribution comes from the low frequency behaviour $q_\sigma^{(a)} \ll 1$ in the case of large N, which is used since the thermodynamic limit $N \to \infty$ is taken, the approximations used in those Eqs. and for the normalisation constant (C.33) are valid, leading to

$$K_D^{(s),ij}(\tau) \cong \frac{C\omega_0}{2\pi} \int_0^\pi dq \, q \cos^2\left(\frac{qD}{2}\right) e^{-\frac{\omega_0 q}{2}\tau}$$

The antisymmetric kernel is treated in the same way, but here the matrix elements $(i = j)$ are shown. With Eqs. (4.21), the case $(i \neq j)$ is easily seen. It reads

$$K_D^{(a),ii}(\tau) \cong \frac{C\omega_0}{2\pi} \int_0^\pi dq \, q \sin^2\left(\frac{qD}{2}\right) e^{-\frac{\omega_0 q}{2}\tau}$$

Now the D-dependence of the matrix elements have to be discussed. Starting with the matrix elements $(i = j)$ the integrals can of course be evaluated exactly and the D dependence disappears by applying a trigonometric identity:

$$K_D^{ii}(\tau) = K_D^{(s),ii}(\tau) + K_D^{(a),ii}(\tau) \cong \frac{C\omega_0}{2\pi} \int_0^\pi dq \, q \underbrace{\left[\cos^2\left(\frac{qD}{2}\right) + \sin^2\left(\frac{qD}{2}\right)\right]}_{=1} e^{-\frac{\omega_0 q}{2}\tau}$$

4.1. TWO DEFECTS

what is left now is the kernel for the case of $i = j$ showing ohmic dissipation with no D-dependence:

$$K_D^{ii}(\tau) \cong \frac{C\omega_0}{2\pi} \int_0^\pi dq\, q\, e^{-\frac{\omega_0 q}{2}\tau} \approx \frac{2C\omega_0}{\pi} \frac{1}{(\omega_0 \tau)^2} \quad (4.22)$$

This makes sense, since D has been chosen in the bulk and the kernel $K_D^{ii}(\tau)$ shows strictly ohmic dissipative behaviour as it should be for an anharmonic bond located in the bulk. The kernel for $i \neq j$ allows to calculate the interaction between both anharmonic bonds with respect to the distance D in between them. Two cases are of interest. One is the distance of both anharmonic bonds D is finite, which should lead to an interaction represented in a non vanishing kernel $K_D^{12}(\tau)$. For the limit of $D \to \infty$ this interaction vanishes as it should be. The only difference to the previously discussed case is a minus sign for the asymmetric kernel leading to a D-dependent result

$$\begin{aligned}
K_D^{i\neq j}(\tau) &= K_D^{(s), i\neq j}(\tau) + K_D^{(a), i\neq j}(\tau) \\
&\cong \frac{C\omega_0}{2\pi} \int_0^\pi dq\, q \left[\cos^2\left(\frac{qD}{2}\right) - \sin^2\left(\frac{qD}{2}\right)\right] e^{-\frac{\omega_0 q}{2}\tau} \quad (4.23)
\end{aligned}$$

As can be seen these kernels ($i \neq j$) have a D-dependence, that needs to be examined carefully. Using a trigonometric identity the cosine and sine squared can be replaced

$$K_D^{i\neq j}(\tau) \cong \frac{C\omega_0}{2\pi} \int_0^\pi dq\, q \cos(qD)\, e^{-\frac{\omega_0 q}{2}\tau} \quad (4.24)$$

, yielding the proof that $\lim_{D \to \infty} K_D^{i\neq j}(\tau) = 0$.
Evaluating this integral (4.24) yields

$$K_D^{i\neq j}(\tau) \cong \frac{C\omega_0}{\pi} \frac{e^{-\frac{\pi\omega_0\tau}{2}} \left(2e^{\frac{\pi\omega_0\tau}{2}}((\omega_0\tau)^2 - 4D^2) - (-1)^D (4D^2(\pi\omega_0\tau - 2) + (\omega_0\tau)^2(2 + \pi\omega_0\tau))\right)}{(4D^2 + (\omega_0\tau)^2)^2}$$

Two q scales occur in the integrand of Eqs. (4.22),(4.22).

$$q_D = \frac{1}{D} \quad \text{and} \quad q_\tau = \frac{1}{\omega_0\tau} \quad (4.25)$$

Equating $q_D = q_\tau$ defines the time scale

$$\tau_D = \frac{D}{\omega_0} \quad (4.26)$$

Considering $\omega_0\tau \gg 1$ one gets from (4.25)

$$K_D^{i\neq j}(\tau) \cong \frac{2C\omega_0}{\pi} \frac{(\omega_0\tau)^2 - 4D^2}{((\omega_0\tau)^2 + 4D^2)^2} \quad (4.27)$$

From this result it is easy to discuss the to relevant cases $\tau \ll \tau_D$ and $\tau \gg \tau_D$.

$$K_D^{i \neq j}(\tau \ll \tau_D) \sim \frac{1}{D^2}$$
$$K_D^{i \neq j}(\tau \gg \tau_D) \cong \frac{2C\omega_0}{\pi} \frac{1}{(\omega_0 \tau)^2} \left[1 - 12 \left(\frac{\tau_D}{\tau}\right)^2\right] \quad (4.28)$$

In the limit $D \to \infty$ the kernel $K_D^{i \neq j}(\tau \ll \tau_D)$ vanishes as it should be, because infinitely large D corresponds to no interaction. For the case of $\tau \gg \tau_D$ it makes no sense taking the limit $D \to \infty$, since τ_D is still considered to be much smaller than τ. The variables q_{M_1} and q_{M_2} and the respective kernels $K_D^{ij}(\tau), i,j = 1,2$ do not correspond to physically relevant quantities, since we are interested in the different (ohmic or super-ohmic) dissipative behaviour that occurs depending on the overall chain-length changing or not. Due to the matrix structure of the kernel a transformation to physically relevant[1] variables has to be considered.

$$q_+ = q_{M_1} + q_{M_2} \quad , \quad q_- = q_{M_1} - q_{M_2} \quad (4.29)$$

Applying this transformation to Eq. (4.19) and using the symmetry relations for the kernel Eq. (4.21) yields

$$S_{infl.}^{E}[q_+, q_-] = -\int_0^T d\tau \int_0^\tau d\tau' \left[K_D^{++}(\tau - \tau')q_+(\tau)q_+(\tau') + K_D^{--}(\tau - \tau')q_-(\tau)q_-(\tau')\right]$$
(4.30)

, with

$$K_D^{++}(\tau) = \frac{K_D^{11}(\tau) + K_D^{12}(\tau)}{2} \overset{\omega_0 \tau \gg 1}{\cong} \begin{cases} \frac{C\omega_0}{\pi}\left[\frac{1}{(\omega_0\tau)^2} - \frac{1}{4D^2}\right] & , \tau \ll \tau_D \\ \frac{2C\omega_0}{\pi} \frac{1}{(\omega_0\tau)^2} & , \tau \gg \tau_D \end{cases}$$

$$K_D^{--}(\tau) = \frac{K_D^{11}(\tau) - K_D^{12}(\tau)}{2} \overset{\omega_0 \tau \gg 1}{\cong} \begin{cases} \frac{C\omega_0}{\pi}\left[\frac{1}{(\omega_0\tau)^2} + \frac{1}{4D^2}\right] & , \tau \ll \tau_D \\ \frac{C\omega_0}{\pi} \frac{24D^2}{(\omega_0\tau)^4} & , \tau \gg \tau_D \end{cases}$$
(4.31)

It is now easy to see, that kernel $K_D^{++}(\tau)$ always shows ohmic dissipative behaviour in leading order, whereas the kernel $K_D^{--}(\tau)$ shows a transition from ohmic to superohmic dissipative behaviour. The kernels $K_D^{++}(\tau), K_D^{--}(\tau)$ are shown in the following figures:

---

[1] the reason for this transformation is given below

4.1. TWO DEFECTS

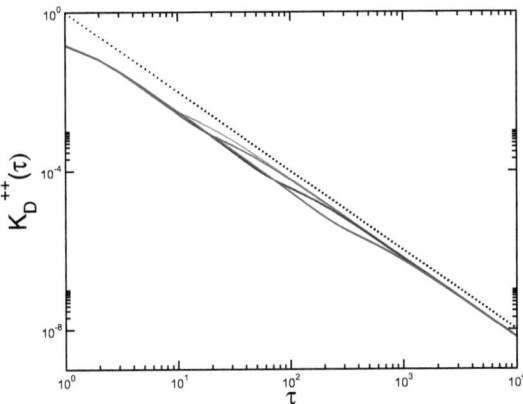

Figure 4.1: τ-dependence of $K_D^{++}(\tau)$ for different D (green: $D = 5$, red: $D = 10$, brown $D = 40$, magenta: $D = 160$) on a log-log representation. The dotted line corresponds to a τ^{-2} behaviour.

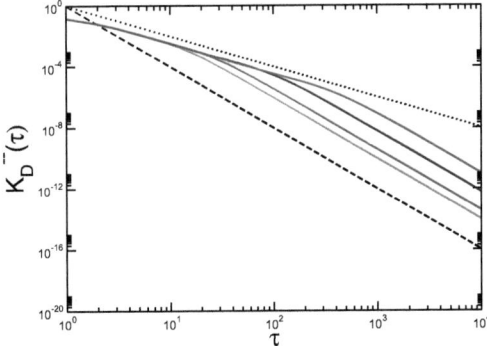

Figure 4.2: τ-dependence of $K_D^{--}(\tau)$ for different D (green: $D = 5$, red: $D = 10$, brown $D = 40$, magenta: $D = 160$) on a log-log representation. The dotted and dashed line, corresponding to τ^{-2} and τ^{-4}, respectively included to see the transition.

In Figure 4.1 the Kernel $K_D^{++}(\tau)$ is shown. It shows a τ^{-2} clearly depicted on the log-log-plot. This plot is a graphical representation of Eq. (4.31), which shows a ohmic dissipative behaviour independent on the introduced timescale τ_D. The small shift resulting from the factor 2 the kernel receives after crossing τ_D is not understood, but nevertheless the dissipative behaviour remains ohmic.

In Figure 4.2 the Kernel $K_D^{--}(\tau)$ is shown. It shows a transition from ohmic dissipative behaviour τ^{-2} to super-ohmic dissipative behaviour τ^{-4}. This transition is clearly visible on the log-log-plot and different anharmonic bond distances D have been included.

Now the advantage of this transformation has to be discussed. In the one anharmonic bond case each of the bonds has two different initial and two different final positions. We were able to show ohmic dissipative behaviour of the Kernel if the anharmonic bond is located in the bulk and a transition from ohmic to super-ohmic dissipative behaviour if the anharmonic bond is located at one of the borders.

For two anharmonic bonds the scenario is more complicated. The reason for switching from ohmic to super-ohmic dissipative behaviour was the position of the anharmonic bond. Tunnelling of the anharmonic bond, located at one of the borders, only requires a finite mass $\sim M = \mathcal{O}(1)$ of the harmonic bath to be moved in the translationally invariant chain. But if the anharmonic bond is located in the bulk, tunnelling of the anharmonic bond requires to move an infinite mass $\sim M = \mathcal{O}(N)$, $N \to \infty$ of the harmonic chain. For two anharmonic bonds located in the bulk, both can tunnel with a movement of an infinite mass or with a movement of a finite mass of the harmonic bath. A movement of a finite mass requires both initial positions of the anharmonic bonds to different, i.e. one having length a_s and the other a_l. Now if one anharmonic bond tunnels and changes its length, the other anharmonic bond can react and also tunnel. The mass that has to be moved is just $\sim D = \mathcal{O}(1)$ and hence finite. Only a finite length $\sim D$ of the harmonic bath had to be moved to allow this tunnelling. Hence one would expect a transition from ohmic to super-ohmic dissipative for times τ depending on the time-scale τ_D in analogy to the one anharmonic bond case, with the bond located at one of the borders.

All other scenarios of both anharmonic bonds tunnelling require a movement of an infinite mass of the harmonic bath, which should result in ohmic dissipative behaviour in analogy to the one anharmonic bond case located in the bulk. Consider for example both anharmonic bonds having lengths a_s, then every tunnelling, no matter if only one anharmonic bond or both bonds tunnel, requires the movement of an infinite mass $\sim \mathcal{O}(N)$, $N \to \infty$, since both bonds are located in the bulk.

The transformation necessary to measure the change in the overall chain-length is given in Eq. (4.29). This is of course only valid for equal equilibrium bond-lengths $a_{s_1} = a_{s_2} \equiv a_s, a_{l_1} = a_{l_2} \equiv a_l$ of the relative coordinates q_{M_1}, q_{M_2}, which is considered here. This is obvious, since if

4.1. TWO DEFECTS

the length-changes of both anharmonic bonds are not the same, movement of an infinite mass of the harmonic chain is still required no matter how the anharmonic bonds tunnel, leading to ohmic dissipative behaviour. The transformation can be interpreted as follows. The kernel matrix $\mathbf{K}_D$, consisting of four elements K_D^{ij} with $i,j = 1,2$ (4.19), is *diagonalised* and contains now only the diagonal elements K_D^{++}, K_D^{--} and no coupling between q_+ and q_- (4.30). The kernel matrix element describing a change in the total length is labeled K_D^{++}, whereas the overall chain-length does not vary for K_D^{--}. Those new coordinates allow to treat the two elements of the influence action of the anharmonic bond case equivalently to the one anharmonic bond case, namely just considering a movement of a finite or an infinite length of the harmonic bath.

- 1-anharmonic bond:

$$\text{anharm. bond at border:} \quad \begin{cases} q_M^- \longrightarrow q_M^+ \\ q_M^+ \longrightarrow q_M^- \end{cases} \quad \text{super-ohmic dissipation}$$

$$\text{anharm. bond in bulk:} \quad \begin{cases} q_M^- \longrightarrow q_M^+ \\ q_M^+ \longrightarrow q_M^- \end{cases} \quad \text{ohmic dissipation}$$

- 2-anharmonic bonds:

$$\text{no overall length-change:} \quad \begin{cases} q_{M_1}^+, q_{M_2}^- \longrightarrow q_{M_1}^-, q_{M_2}^+ \\ q_{M_1}^-, q_{M_2}^+ \longrightarrow q_{M_1}^+, q_{M_2}^- \end{cases} \quad \text{super-ohmic dissipation}$$

$$\text{big overall length-change:} \quad \begin{cases} q_{M_1}^-, q_{M_2}^- \longrightarrow q_{M_1}^+, q_{M_2}^+ \\ q_{M_1}^+, q_{M_2}^+ \longrightarrow q_{M_1}^-, q_{M_2}^- \end{cases} \quad \text{ohmic dissipation}$$

$$\text{small overall length-change:} \quad \begin{cases} q_{M_1}^-, q_{M_2}^- \longrightarrow q_{M_1}^-, q_{M_2}^+ \\ q_{M_1}^-, q_{M_2}^- \longrightarrow q_{M_1}^+, q_{M_2}^- \\ q_{M_1}^+, q_{M_2}^+ \longrightarrow q_{M_1}^-, q_{M_2}^+ \\ q_{M_1}^+, q_{M_2}^+ \longrightarrow q_{M_1}^+, q_{M_2}^- \\ q_{M_1}^-, q_{M_2}^+ \longrightarrow q_{M_1}^-, q_{M_2}^- \\ q_{M_1}^-, q_{M_2}^+ \longrightarrow q_{M_1}^+, q_{M_2}^+ \\ q_{M_1}^+, q_{M_2}^- \longrightarrow q_{M_1}^+, q_{M_2}^+ \\ q_{M_1}^+, q_{M_2}^- \longrightarrow q_{M_1}^-, q_{M_2}^- \end{cases} \quad \text{ohmic dissipation}$$

The case of small overall length-change requires of course tunnelling of the anharmonic bonds at different times. This graphic shows, that even though there are many more paths in the two anharmonic bond case compared to the one anharmonic bond case, the number of different kernels describing the paths does not change. In the one anharmonic bond case the kernel was either ohmic or super-ohmic dependent on the position of the anharmonic bond, whereas in the case of two anharmonic bonds the kernel is also either ohmic or super-ohmic, dependent on the initial and final positions of the anharmonic bonds. The transformation from q_{M_1}, q_{M_2} to q_+, q_- does not only decouple the influence action, but introduces a new coupling in the local action, namely a term $\sim q_+^2 q_-^2$. This term yields a coupling of the up to now bare instantons q_1, q_2 in the local action. The effect of this coupling is not fully understood.

4.2 Tunnelling expectation value using extended NIBA

Defining a measurable quantity $p(t) = \text{Tr}\left[\rho_{red}(t)\sigma_z^{(1)} \otimes \sigma_z^{(2)}\right]$, where the matrix elements of the reduced density operator $\rho_{red}(t)$ have been calculated Appendix E and are given in form of a path integral Eq. (E.14). The derivation of the influence functional $\mathcal{F}$ for the two anharmonic bond case is similar to the derivation of the one anharmonic bond case. The derivation is given in detail in Appendix E. Using the influence functional for two anharmonic bonds

$$\mathcal{F}[q_{M_1}, q_{M_2}, q'_{M_1}, q'_{M_2}] = \exp\left[-\frac{1}{\hbar}\sum_{a,b=1}^{2}\int_0^t d\tau \int_0^\tau d\tau' \left(q_{M_a}(\tau) - q'_{M_a}(\tau)\right) \right.$$
$$\left. \cdot \left(L_{ab}(\tau - \tau')q_{M_b}(\tau') - L_{ab}^*(\tau - \tau')q'_{M_b}(\tau')\right)\right] \quad (4.32)$$

where the function $L_{ab}(\tau)$ is defined, for zero temperature, as

$$L_{ab}(\tau) = \sum_{\sigma=1}^{N-3} \frac{c_{a,\sigma} c_{b,\sigma}}{2m_\sigma \omega_\sigma} \left[\cos(\omega_\sigma \tau) - i\sin(\omega_\sigma \tau)\right] \quad (4.33)$$

Splitting the function $L_{ab}(\tau)$ into a real and an imaginary part one can put this influence functional in the following form

$$\mathcal{F}[\xi^{(1)}(\tau), \chi^{(1)}(\tau); \xi^{(2)}(\tau'), \chi^{(2)}(\tau')] =$$
$$\exp\left[-\frac{q_0^2}{\pi\hbar}\sum_{a,b=1}^{2}\int_0^t d\tau \int_0^\tau d\tau' \left(L_2^{ab}(\tau-\tau')\xi^{(a)}(\tau)\xi^{(b)}(\tau') - iL_1^{ab}(\tau-\tau')\xi^{(a)}(\tau)\chi^{(b)}(\tau')\right)\right]$$
$$(4.34)$$

by using

$$L_1^{ab}(\tau - \tau') = \int_0^\infty d\omega \, J_{ab}(\omega) \sin\left(\omega[\tau - \tau']\right)$$
$$L_2^{ab}(\tau - \tau') = \int_0^\infty d\omega \, J_{ab}(\omega) \cos\left(\omega[\tau - \tau']\right) \quad (4.35)$$

and applying (as in the one anharmonic bond case) a transformation of the coordinates q_{M_a}, q'_{M_a} to blips ξ_a and sojourns χ_a with the following transformation

$$\xi^{(a)}(\tau) = \frac{q_{M_a}(\tau) - q'_{M_a}(\tau)}{q_0}$$
$$\chi^{(a)}(\tau) = \frac{q_{M_a}(\tau) + q'_{M_a}(\tau)}{q_0} \quad (4.36)$$

The spectral density $J_{ab}(\omega)$ is defined as

$$J_{ab}(\omega) = \frac{\pi}{2} \sum_{\sigma=1}^{N-3} \frac{c_{a,\sigma} c_{b,\sigma}}{m_\sigma \omega_\sigma} \delta(\omega - \omega_\sigma) \qquad (4.37)$$

Splitting the spectral density into a symmetric (s) and an antisymmetric (a) part and using the definitions of the variables from Eqs. (4.16), (4.17), one gets

$$\begin{aligned}J_{aa}(\omega) &= \frac{C\omega_0 \pi}{2} \sum_{\sigma=1}^{\frac{N}{2}-1} \mathcal{N}_{\sigma,s}^2 \sin^2\left(q_\sigma^{(s)} \frac{N-D}{2}\right) \sin\left(\frac{q_\sigma^{(s)}}{2}\right) \delta\left(\omega - \omega_0 \sin\left(\frac{q_\sigma^{(s)}}{2}\right)\right) \\ &+ \frac{C\omega_0 \pi}{2} \sum_{\sigma=1}^{\frac{N}{2}-2} \mathcal{N}_{\sigma,a}^2 \sin^2\left(q_\sigma^{(a)} \frac{N-D}{2}\right) \sin\left(\frac{q_\sigma^{(a)}}{2}\right) \delta\left(\omega - \omega_0 \sin\left(\frac{q_\sigma^{(a)}}{2}\right)\right) \end{aligned}$$
(4.38)

Replacing the normalisation constants for finite $D = \mathcal{O}(1)$ according to Eq. (C.33), splitting the squared sine into one component containing N and the other containing D, we are able to get rid of the N dependence. To do this, the results from the transcendental equation Eqs. (C.30) are used. We are then able to perform the thermodynamic limit ($N \to \infty$), yielding

$$\begin{aligned}J_{aa}(\omega) &= \frac{C\omega_0}{2} \int_0^\pi dq \, \sin\left(\frac{q}{2}\right) \delta\left(\omega - \omega_0 \sin\left(\frac{q}{2}\right)\right) \\ &\cdot \left[\frac{1}{1+f_s^2(q,D)} \left\{ \cos^2\left(\frac{qD}{2}\right) f_s^2(q,D) - \sin(qD) f_s(q,D) + \sin^2\left(\frac{qD}{2}\right) \right\} \right. \\ &\left. + \frac{1}{1+f_a^2(q,D)} \left\{ \cos^2\left(\frac{qD}{2}\right) - \sin(qD) f_a(q,D) + \sin^2\left(\frac{qD}{2}\right) f_a^2(q,D) \right\} \right]\end{aligned}$$
(4.39)

Evaluating the integral gives

$$\begin{aligned}J_{aa}(\omega) &= \frac{C\omega}{\omega_0} \left(1 - \left(\frac{\omega}{\omega_0}\right)^2\right)^{-\frac{1}{2}} \left(\frac{1}{1+f_s^2\left(\frac{\omega}{\omega_0},D\right)} \left[\cos^2\left(D \arcsin\left(\frac{\omega}{\omega_0}\right)\right) f_s^2\left(\frac{\omega}{\omega_0},D\right) \right.\right. \\ &\left.- \sin\left(2D \arcsin\left(\frac{\omega}{\omega_0}\right)\right) f_s\left(\frac{\omega}{\omega_0},D\right) + \sin^2\left(D \arcsin\left(\frac{\omega}{\omega_0}\right)\right) \right] \\ &+ \frac{1}{1+f_a^2\left(\frac{\omega}{\omega_0},D\right)} \left[\cos^2\left(D \arcsin\left(\frac{\omega}{\omega_0}\right)\right) \right. \\ &\left.\left.- \sin\left(2D \arcsin\left(\frac{\omega}{\omega_0}\right)\right) f_a\left(\frac{\omega}{\omega_0},D\right) + \sin^2\left(D \arcsin\left(\frac{\omega}{\omega_0}\right)\right) f_a^2\left(\frac{\omega}{\omega_0},D\right) \right] \right)\end{aligned}$$
(4.40)

4.2. TUNNELLING EXPECTATION VALUE USING EXTENDED NIBA

Since the quantum dissipation generated by the harmonic bath depends on the low frequencies $\omega \ll \omega_0$ of the spectral density, it is possible to Taylor expand the result, considering finite D, yielding

$$J_{aa}(\omega \ll \omega_0) = C\frac{\omega}{\omega_0} - C\frac{7+8D}{2}\left(\frac{\omega}{\omega_0}\right)^3 + \mathcal{O}\left(\left(\frac{\omega}{\omega_0}\right)^5\right) \quad (4.41)$$

The calculation of $J_{a\neq b}(\omega)$ is done analogously. The result is

$$J_{a\neq b}(\omega \ll \omega_0) = C\frac{\omega}{\omega_0} - C\frac{7+8D+4D^2}{2}\left(\frac{\omega}{\omega_0}\right)^3 + \mathcal{O}\left(\left(\frac{\omega}{\omega_0}\right)^5\right) \quad (4.42)$$

As in the one anharmonic bond case a transformation of the blip and sojourn variables to "charges" is performed, but one essential assumption is included in the two anharmonic bond case. Since it seems impossible to calculate the tunnelling probability of two anharmonic bonds tunnelling at different timesteps, a crude assumption, namely setting the timesteps of both anharmonic bonds equal, is made.

$$
\begin{aligned}
\xi^{(1)}(\tau) &= \sum_{j=1}^{n}\zeta_j^{(1)}\left[\Theta(\tau-t_{2j-1})-\Theta(\tau-t_{2j})\right] \\
\xi^{(2)}(\tau) &= \sum_{j=1}^{n}\zeta_j^{(2)}\left[\Theta(\tau-t_{2j-1})-\Theta(\tau-t_{2j})\right] \\
\chi^{(1)}(\tau) &= \sum_{j=0}^{n}\eta_j^{(1)}\left[\Theta(\tau-t_{2j})-\Theta(\tau-t_{2j+1})\right] \\
\chi^{(2)}(\tau) &= \sum_{j=0}^{n}\eta_j^{(2)}\left[\Theta(\tau-t_{2j})-\Theta(\tau-t_{2j+1})\right]
\end{aligned} \quad (4.43)
$$

This approximation forces one of the anharmonic bonds to react instantaneously to a tunnelling of one of the other bond variables. This is of course not the case, since both anharmonic bonds are separated by the distance D, but the time information needs to travel from one anharmonic bond to the other is given by the time a phonon needs $t_D = \omega_0 D$ to cross this distance. But for small distances D the approximation gets better and better. The detailed calculation of the above (4.43) transformation applied to the influence functional (4.34) is done in Appendix D. Equating the tunnelling of both anharmonic bonds has also been applied by [29]. They write "when the indirect coupling is the largest energy scale, the two spins will tend to tunnel simultaneously". Since there is no direct coupling between the two anharmonic bonds, but only the indirect coupling through the harmonic bath, this assumption is always fulfilled. The only difference, that the two continuous anharmonic bond coordinates are no spins, but are considered to flip instantaneously, which makes it possible to treat them as spins. In [35] it is written, that "For the double-impurity case, ..., a low-temperature and short-distance regime, where correlated tunnelling is established ...", which also strengthens the assumption used in this thesis of setting the flipping times of both anharmonic bonds equal under the restrictions given above.

Another point shows the validity of this approximation. The influence functional Eq. (4.32) has an oscillating part containing the imaginary part of the function $L^{ab}(t)$ in the exponent. This term is called friction term. The other term, with the real part of the function $L^{ab}(t)$ in the exponent, is called the noise term. This noise term is always bigger than zero and randomly pumps back energy to the system. In the one anharmonic bond case, summing over the same subset of blip-charges, makes this term always positive, as mentioned by Leggett et al. [14]. But for two anharmonic bonds a new situation arises. The argument of the exponent of the noise term looks like

$$-\frac{q_0^2}{\pi\hbar}\int_0^t d\tau \int_0^\tau d\tau' \left[L_2^{11}(\tau-\tau')\xi^{(1)}(\tau)\xi^{(1)}(\tau') + L_2^{12}(\tau-\tau')\xi^{(1)}(\tau)\xi^{(2)}(\tau') \right.$$
$$\left. + L_2^{21}(\tau-\tau')\xi^{(2)}(\tau)\xi^{(1)}(\tau') + L_2^{22}(\tau-\tau')\xi^{(2)}(\tau)\xi^{(2)}(\tau') \right] \qquad (4.44)$$

Now one has to look at all possible cases, they are

1) $\xi^{(1)} = \xi^{(2)} = \begin{cases} +1, & a \\ -1, & b \\ 0, & c \end{cases}$

2) $\xi^{(1)} \neq \xi^{(2)} = \begin{cases} (+1,-1), & a \\ (-1,+1), & b \\ (+1,0), & c \\ (0,+1), & d \\ (-1,0), & e \\ (0,-1), & f \end{cases}$

Using all possible cases in (4.44), one gets for the argument

- 0,   1c

- $-\frac{2q_0^2}{\pi\hbar}Q_2^-(t)$,   2a, 2b

- $-\frac{q_0^2}{\pi\hbar}Q_2(t)$,   2c, 2d, 2e, 2f

4.2. TUNNELLING EXPECTATION VALUE USING EXTENDED NIBA

- $-\frac{2q_0^2}{\pi\hbar}Q_2^+(t)$,   1a, 1b

, with

$$Q_{1/2}^{(aa)}(t) \equiv Q_{1/2}(t), \qquad a = 1, 2$$
$$Q_{1/2}^{\pm}(t) = Q_{1/2}(t) \pm Q_{1/2}^{(12)}(t) \tag{4.45}$$

, which implies of course also

$$J^{\pm}(\omega) = J_{aa}(\omega) \pm J_{a\neq b}(\omega) \tag{4.46}$$

The functions $Q_{1/2}^{(ab)}$ where defined in (4.75) and are the result of performing the two time integrations $\int_0^t d\tau \int_0^\tau d\tau'$ of the functions $L_1^{ab}(\tau - \tau'), L_2^{ab}(\tau - \tau')$ in Eq. (4.44).

The possible cases are given in descending order, largest value on top smallest value at the bottom. This can be seen in the interesting[2] time regime $\omega_0 t \gg 1$[3] in Eq. (4.86), where $Q_2(t)$(4.48) has of course the same time dependence in the long time limit as $Q_2^+(t)$(4.86). Using the calculation of the spectral density $J_{aa}(\omega)$(4.41) it is possible to calculate $Q_2(t)$

$$Q_2(t) = \frac{C}{2\omega_0}\left[\frac{(7+8D)(\cos(\omega_0 t)-1)}{(\omega_0 t)^2} + \frac{(7+8D)\sin(\omega_0 t)}{\omega_0 t} + 2\ln(\omega_0 t) - 2\text{Ci}(\omega_0 t)\right] \tag{4.47}$$

now considering the long time limit $\omega_0 t \gg 1$ one gets

$$Q_2(t) = \frac{C}{\omega_0}\ln(\omega_0 t) \tag{4.48}$$

The noise term is maximised for the case of 1c, which is nothing but both anharmonic bonds being at the same time in a sojourn state. This is physically reasonable, since maximising the noise term in the one anharmonic bond case acts as a Gaussian filter quenching off-diagonal quantum fluctuations [16]. It is also the reason, why the NIBA is a valid approximation in the one anharmonic bond case, since the system stays longer in a sojourn state and hence the blips can be treated as a dilute gas.

This is already enough to assume, that both anharmonic bonds tend two be either together in a sojourn state or together in a blip state, rather then one in a sojourn and the other in a blip, at the same time.

The second biggest value for the noise functional is achieved in the case of 2a, 2b, which is both anharmonic bonds being at the same time in different blip states. This clearly shows,

---

[2]As has been mentioned in [14], for $\omega_c t \lesssim 1$ the corrections to 1 in $P_1(t)$ (and to 0 in $P_{2/3}(t)$ respectively), are of order $\left(\frac{\Delta}{\omega_c}\right)^2$, at most and hence of little interest.

[3]Since we assume $t \gg t_D = \frac{D}{\omega_0}$, the case of $\omega_0 t \gg 1$ is of course always fulfilled since $D \geq 2$.

that both anharmonic bonds tend to be either both in a sojourn state or both in different blip states, compared to one being in a blip and the other one being in a sojourn state. The only case even less probable than one bond in a sojourn and the other in a blip state, is both bonds being in the same blip state 1a, 1b.

This effect, that both bonds tend to be at the same time in sojourn states or different blip states, is captured by forcing both anharmonic bonds to behave as one bond. Since we are investigating both anharmonic bonds changing the states of their tunnelling coordinates $q_{M_1}, q'_{M_1}, q_{M_2}, q'_{M_2}$ at the same timesteps, it is possible to treat the two two-state coordinates (every coordinate has two possible values $\pm\frac{q_0}{2}$) as one four state coordinate. The total bare amplitudes $A[q_{M_1}(\tau)], B[q_{M_2}(\tau)]$ for a given path are now $A[q_{M_1}(\tau), q_{M_2}(\tau)]$ and may be broken up into small pieces of length dt[14]. The amplitude to remain in the state is one, whereas the amplitude of switching states is $i\frac{\Delta}{2}dt$. In the one anharmonic bond case [14] the basis for defining the tunnelling amplitudes is the spin-boson Hamiltonian without bias ϵ and no coupling to the bath Eq. (1.1) [14].

$$H_{\epsilon=0} = -\frac{\hbar\Delta}{2}\sigma_x = -\frac{\hbar\Delta}{2}\begin{pmatrix} 0 & 1 \\ 1 & 0 \end{pmatrix}$$

The basis is formed by the localized states $|R\rangle = \begin{pmatrix} 1 \\ 0 \end{pmatrix}, |L\rangle = \begin{pmatrix} 0 \\ 1 \end{pmatrix}$ representing the right or left well of the symmetric double well potential $V(q)$, which are eigenstates of σ_z belonging to the eigenvalues $1, -1$, respectively. The path integral of the bare amplitudes defined in [14] Eq. (4.1) is nothing but the tunnelling matrix element of the time-evolution of the above given Hamiltonian.

$$\int_{q_i}^{q_f} \mathcal{D}q(\tau)A[q(\tau)] = \langle q_f|e^{-\frac{iH_{\epsilon=0}t}{\hbar}}|q_i\rangle$$

$$\int_{q'_i}^{q'_f} \mathcal{D}q'(\tau)A^*[q'(\tau')] = \langle q'_f|e^{\frac{iH_{\epsilon=0}t}{\hbar}}|q'_i\rangle \quad (4.49)$$

In the following we will only calculate the bare tunnelling amplitude of $q(\tau)$, the results for $q'(\tau')$ are just the complex conjugate results. Splitting the time into small time steps dt allows the Taylor expansion of the exponential leading to

$$\langle q_f|e^{-\frac{iH_{\epsilon=0}dt}{\hbar}}|q_i\rangle = \langle q_f|q_i\rangle - \frac{i\,dt}{\hbar}\langle q_f|H_{\epsilon=0}|q_i\rangle + \mathcal{O}(dt^2) \quad (4.50)$$

4.2. TUNNELLING EXPECTATION VALUE USING EXTENDED NIBA 61

Two examples are explicitely calculated, up to $\mathcal{O}(dt^2)$, to show how the tunnelling matrix element is defined

$$\langle R|e^{-\frac{iH_{\epsilon=0}dt}{\hbar}}|L\rangle = \underbrace{\langle R|L\rangle}_{=0} - \frac{i\,dt}{\hbar}\langle R|H_{\epsilon=0}|L\rangle + \mathcal{O}(dt^2)$$

$$= \frac{i\Delta}{2}dt \begin{pmatrix} 1 \\ 0 \end{pmatrix} \begin{pmatrix} 0 & 1 \\ 1 & 0 \end{pmatrix} \begin{pmatrix} 0 \\ 1 \end{pmatrix} + \mathcal{O}(dt^2)$$

$$= \frac{i\Delta}{2}dt + \mathcal{O}(dt^2) \tag{4.51}$$

$$\langle R|e^{-\frac{iH_{\epsilon=0}dt}{\hbar}}|R\rangle = \underbrace{\langle R|R\rangle}_{=1} - \frac{i\,dt}{\hbar}\langle R|H_{\epsilon=0}|R\rangle + \mathcal{O}(dt^2)$$

$$= 1 + \frac{i\Delta}{2}dt \begin{pmatrix} 1 \\ 0 \end{pmatrix} \begin{pmatrix} 0 & 1 \\ 1 & 0 \end{pmatrix} \begin{pmatrix} 1 \\ 0 \end{pmatrix} + \mathcal{O}(dt^2)$$

$$= 1 + \mathcal{O}(dt^2) \tag{4.52}$$

This is how the path integral of the bare tunnelling amplitude for the spin-boson Hamiltonian without bias ϵ is calculated in [14]. The amplitude $A^*[q'(\tau')]$ leads to the same results, except for a minus in the switching amplitude $\frac{i\Delta}{2}dt$. The two-state variables $q(\tau), q'(\tau')$ have been considered as a pair $[q(\tau), q'(\tau')]$ jumping between four states[14]. The four possible states are

$$\begin{aligned} A &= \{+,+\} \\ B &= \{+,-\} \\ C &= \{-,+\} \\ D &= \{-,-\} \end{aligned} \tag{4.53}$$

, where $+ \equiv +\frac{q_0}{2}, - \equiv -\frac{q_0}{2}$. The tunnelling amplitudes, up to $\mathcal{O}(dt^2)$, are hence

$$\begin{aligned} 0 &\begin{cases} A \leftrightarrow D \\ B \leftrightarrow C \end{cases} \\ -i\frac{\Delta}{2}dt &\begin{cases} A \leftrightarrow B \\ D \leftrightarrow C \end{cases} \\ i\frac{\Delta}{2}dt &\begin{cases} A \leftrightarrow C \\ D \leftrightarrow B \end{cases} \end{aligned} \tag{4.54}$$

For the two anharmonic bond case, where both bonds tunnel at the same time both bonds can be considered as one *super-bond*.
The amplitudes $A[q_{M_1}(\tau)]B[q_{M_2}(\tau)], A^*[q'_{M_1}(\tau')]B^*[q'_{M_2}(\tau')]$ to switch between states during the

time intervall dt merge together to $A[q_{M_1}(\tau), q_{M_2}(\tau)]$, $A^*[q'_{M_1}(\tau'), q'_{M_2}(\tau')]$ and are defined analogously to the one anharmonic bond case as

$$-i\frac{\Delta}{2}dt \begin{cases} AA \leftrightarrow BB \\ AD \leftrightarrow BC \\ DA \leftrightarrow CB \\ DD \leftrightarrow CC \end{cases}$$

$$i\frac{\Delta}{2}dt \begin{cases} AA \leftrightarrow CC \\ AD \leftrightarrow BC \\ DA \leftrightarrow CB \\ DD \leftrightarrow BB \end{cases}$$

all others are 0 \hfill (4.55)

Where the states are defined analogeously to the one anharmonic bond case in the following way

$$\begin{aligned}
AA &= \{++,++\} \\
AB &= \{++,+-\} \\
AC &= \{++,-+\} \\
AD &= \{++,--\} \\
BA &= \{+-,++\} \\
BB &= \{+-,+-\} \\
BC &= \{+-,-+\} \\
BD &= \{+-,--\} \\
CA &= \{-+,++\} \\
CB &= \{-+,+-\} \\
CC &= \{-+,-+\} \\
CD &= \{-+,--\} \\
DA &= \{--,++\} \\
DB &= \{--,+-\} \\
DC &= \{--,-+\} \\
DD &= \{--,--\}
\end{aligned} \qquad (4.56)$$

4.2. TUNNELLING EXPECTATION VALUE USING EXTENDED NIBA

The calculation of these tunnelling amplitudes is done analogeously to the one anharmonic case, considering a spin-boson Hamiltonian of the following form

$$H_{\epsilon=0} = -\frac{\hbar\Delta}{2}\sigma_x^{(1)} \otimes \sigma_x^{(2)} = -\frac{\hbar\Delta}{2}\begin{pmatrix} 0 & 0 & 0 & 1 \\ 0 & 0 & 1 & 0 \\ 0 & 1 & 0 & 0 \\ 1 & 0 & 0 & 0 \end{pmatrix} \quad (4.57)$$

The basis is formed by the localized states

$$|RR\rangle = |R\rangle \otimes |R\rangle = \begin{pmatrix} 1 \\ 0 \\ 0 \\ 0 \end{pmatrix}$$

$$|RL\rangle = |R\rangle \otimes |L\rangle = \begin{pmatrix} 0 \\ 1 \\ 0 \\ 0 \end{pmatrix}$$

$$|LR\rangle = |L\rangle \otimes |R\rangle = \begin{pmatrix} 0 \\ 0 \\ 1 \\ 0 \end{pmatrix}$$

$$|LL\rangle = |L\rangle \otimes |L\rangle = \begin{pmatrix} 0 \\ 0 \\ 0 \\ 1 \end{pmatrix} \quad (4.58)$$

representing the four possible states, which are eigenstates of $\sigma_z^{(1)} \otimes \sigma_z^{(2)}$ belonging to the two-fold degenerate eigenvalues $1, -1$, respectively. Now following the same procedure as in the one anharmonic bond case three possible transitions are calculated in detail to prove the assumption of Eq.(4.55).

$$\langle RR|e^{-\frac{iH_{\epsilon=0}dt}{\hbar}}|RL\rangle = \underbrace{\langle RR|RL\rangle}_{=0} - \frac{i\,dt}{\hbar}\langle RR|H_{\epsilon=0}|RL\rangle + \mathcal{O}(dt^2)$$

$$= \frac{i\Delta}{2}dt \begin{pmatrix} 1 \\ 0 \\ 0 \\ 0 \end{pmatrix} \begin{pmatrix} 0 & 0 & 0 & 1 \\ 0 & 0 & 1 & 0 \\ 0 & 1 & 0 & 0 \\ 1 & 0 & 0 & 0 \end{pmatrix} \begin{pmatrix} 0 \\ 1 \\ 0 \\ 0 \end{pmatrix} + \mathcal{O}(dt^2)$$

$$= 0 + \mathcal{O}(dt^2) \qquad (4.59)$$

$$\langle LL|e^{-\frac{iH_{\epsilon=0}dt}{\hbar}}|RR\rangle = \underbrace{\langle LL|RR\rangle}_{=0} - \frac{i\,dt}{\hbar}\langle RR|H_{\epsilon=0}|LL\rangle + \mathcal{O}(dt^2)$$

$$= \frac{i\Delta}{2}dt \begin{pmatrix} 1 \\ 0 \\ 0 \\ 0 \end{pmatrix} \begin{pmatrix} 0 & 0 & 0 & 1 \\ 0 & 0 & 1 & 0 \\ 0 & 1 & 0 & 0 \\ 1 & 0 & 0 & 0 \end{pmatrix} \begin{pmatrix} 0 \\ 0 \\ 0 \\ 1 \end{pmatrix} + \mathcal{O}(dt^2)$$

$$= \frac{i\Delta}{2}dt + \mathcal{O}(dt^2) \qquad (4.60)$$

$$\langle LR|e^{-\frac{iH_{\epsilon=0}dt}{\hbar}}|RL\rangle = \underbrace{\langle LR|RL\rangle}_{=0} - \frac{i\,dt}{\hbar}\langle LR|H_{\epsilon=0}|RL\rangle + \mathcal{O}(dt^2)$$

$$= \frac{i\Delta}{2}dt \begin{pmatrix} 0 \\ 0 \\ 1 \\ 0 \end{pmatrix} \begin{pmatrix} 0 & 0 & 0 & 1 \\ 0 & 0 & 1 & 0 \\ 0 & 1 & 0 & 0 \\ 1 & 0 & 0 & 0 \end{pmatrix} \begin{pmatrix} 0 \\ 1 \\ 0 \\ 0 \end{pmatrix} + \mathcal{O}(dt^2)$$

$$= \frac{i\Delta}{2}dt + \mathcal{O}(dt^2) \qquad (4.61)$$

With the bare tunnelling amplitude $A^*[q'_{M_1}(\tau'), q'_{M_2}(\tau')]$ leading to the same results, except for the minus in the tunnelling amplitude $\frac{i\Delta}{2}dt$ as in the one anharmonic bond case, it is now possible to calculate the tunnelling amplitudes and verify the assumption of Eq. (4.55). Since both anharmonic bonds tunnel at the same time, states such as e.g. $AB, AC, BD, \ldots$ are not allowed and do not occur due to the spin-boson Hamiltonian (4.57), because the starting state **must** be AA, AD, DA or DD. This means $q_{M_1} = q'_{M_1}$ and $q_{M_2} = q'_{M_2}$, which is absolutely equivalent to the requirement of starting in state A or D in the one anharmonic bond case considered by [14]. The approximation of forcing both anharmonic bonds to tunnel at the same time is one of the main differences to the scenario presented by [29]. This thesis restricts already the bare tunnelling amplitudes and hence investigates a $P(t)$, where both bonds tunnel at the same timesteps dt. Dubé and Stamp derive a $P(t)$ where both bonds can tunnel at different timesteps dt, du and discuss certain special scenarios, where the timescales are set equal. The

4.2. TUNNELLING EXPECTATION VALUE USING EXTENDED NIBA

mathematical rigor used by Dubé and Stamp in setting the timescales equal is questionable, since there are no explicit calculations given and most of the approximations are hand-waving arguments. The probability $p(t)$ derived in Appendix D Eq. (E.20) will be presented for four different cases (labelled $p_1(t), p_2(t), p_3(t), p_4(t)$), depending on the relation of the initial and final states. The probability $p(t)$ for starting and ending in the same state is for all possible initial states

$$p_{1/2}(t) = 1 + \frac{1}{2}\sum_{n=1}^{\infty}(-1)^n \Delta^{2n} K_n^{(1/2)}(t) \qquad (4.62)$$

, whereas ending in a different state from the one starting from, gives

$$p_{3/4}(t) = 1 - p_{1/2}(t) \qquad (4.63)$$

The indices 1/2 distinguish the following scenarios

initial **and** final state for index 1: $\quad q_{M_1} = q'_{M_1} = q_{M_2} = q'_{M_2}$
, which corresponds to the choice of AA or DD

initial **and** final state for index 2: $\quad q_{M_1} = q'_{M_1} \neq q_{M_2} = q'_{M_2}$
, which corresponds to the choice of AD or DA $\qquad (4.64)$

, whereas the indices 3/4 label the scenarios presented below

initial **not equal** final state for index 3: $\quad q_{M_1} = q'_{M_1} = q_{M_2} = q'_{M_2}$
, which corresponds to the choice of AA or DD

initial **not equal** final state for index 4: $\quad q_{M_1} = q'_{M_1} \neq q_{M_2} = q'_{M_2}$
, which corresponds to the choice of AD or DA $\qquad (4.65)$

Because of Eq. (4.63) it is only needed to investigate $p_{1/2}(t)$, since it already contains all the information about the other two cases, as has already been noticed in [14] for the case of one anharmonic bond.

The effect of only allowing tunneling of both anharmonic bonds at equal timesteps can be investigated best by looking at Eq. (4.34). Lets consider $p_1(t)$ for the choice of the initial states $q_{M_1}(0) = q'_{M_1}(0) = q_{M_2}(0) = q'_{M_2}(0) = +\frac{q_0}{2}$. Since the tunnelling is restricted to $q_{M_1}(\tau), q_{M_2}(\tau)$ and respectively $q'_{M_1}(\tau), q'_{M_2}(\tau)$ tunnelling at the same time, the choice of the initial states and the free choice of the path of one variable (here $q_{M_1}(\tau)$) defines **all** the other pathes. For the initial states chosen above and the restriction of tunnelling at the same time, the following relations hold as can be easily verified

$$\begin{aligned}\xi^{(1)}(\tau) &= \xi^{(2)}(\tau) \\ \chi^{(1)}(\tau) &= \chi^{(2)}(\tau)\end{aligned} \qquad (4.66)$$

The initial positions of the anharmonic bonds are defined above and yield the following functions with use of Eq. (4.36)

$$\begin{aligned}
\xi^{(1)}(0) &= \frac{q_{M_1}(0) - q'_{M_1}(0)}{q_0} = 0 = \xi^{(2)}(0) = \frac{q_{M_2}(0) - q'_{M_2}(0)}{q_0} \\
\chi^{(1)}(0) &= \frac{q_{M_1}(0) + q'_{M_1}(0)}{q_0} = 1 = \chi^{(2)}(0) = \frac{q_{M_2}(0) + q'_{M_2}(0)}{q_0}
\end{aligned} \quad (4.67)$$

Now the first tunnelling process happens at timestep t_1. Lets have $q_{M_1}(\tau = t_1)$ switch states from $+\frac{q_0}{2}$ to $-\frac{q_0}{2}$ and see what happens. Since we forced both bonds to tunnel at the same time also $q_{M_2}(\tau = t_1)$ switches its state from $+\frac{q_0}{2}$ to $-\frac{q_0}{2}$. As in the publication [14] either q or q' switches, so $q'_{M_1}(\tau = t_1), q'_{M_2}(\tau = t_1)$ stay as they were. This yields

$$\begin{aligned}
\xi^{(1)}(\tau = t_1) &= \frac{q_{M_1}(\tau = t_1) - q'_{M_1}(\tau = t_1)}{q_0} = -1 = \xi^{(2)}(\tau = t_1) = \frac{q_{M_2}(\tau = t_1) - q'_{M_2}(\tau = t_1)}{q_0} \\
\chi^{(1)}(\tau = t_1) &= \frac{q_{M_1}(\tau = t_1) + q'_{M_1}(\tau = t_1)}{q_0} = 0 = \chi^{(2)}(\tau = t_1) = \frac{q_{M_2}(\tau = t_1) + q'_{M_2}(\tau = t_1)}{q_0}
\end{aligned}$$
(4.68)

Now the second tunnelling process happens at timestep t_2. Here we let $q'_{M_1}(\tau = t_2)$ switch states from $+\frac{q_0}{2}$ to $-\frac{q_0}{2}$. As in the case before $q'_{M_2}(\tau = t_1)$ has to switch its state from $+\frac{q_0}{2}$ to $-\frac{q_0}{2}$, whereas $q_{M_1}(\tau = t_2), q_{M_2}(\tau = t_2)$ stay as they were, yielding

$$\begin{aligned}
\xi^{(1)}(\tau = t_2) &= \frac{q_{M_1}(\tau = t_2) - q'_{M_1}(\tau = t_2)}{q_0} = 0 = \xi^{(2)}(\tau = t_2) = \frac{q_{M_2}(\tau = t_2) - q'_{M_2}(\tau = t_2)}{q_0} \\
\chi^{(1)}(\tau = t_2) &= \frac{q_{M_1}(\tau = t_2) + q'_{M_1}(\tau = t_2)}{q_0} = -1 = \chi^{(2)}(\tau = t_2) = \frac{q_{M_2}(\tau = t_2) + q'_{M_2}(\tau = t_2)}{q_0}
\end{aligned}$$
(4.69)

Now we look at the flips that occured. Initially ($\tau = 0$) we started with AA, the first tunnelling at $\tau = t_1$ switches to CC and the second tunnelling at $\tau = t_2$ switches to DD. What we see and can easily be calculated is, that the initial states AA, DD allow only tunnelling to the states BB, CC. The other types of initial states AD, DA allow only tunnelling to the states BC, CB as has already been defined in Eq. (4.55). It is obvious now what kind of tunnelling processes are described by $p_i(t)$, $i = 1, 2, 3, 4$.

As mentioned above we will only look at $p_i(t)$, $i = 1, 2$. The calculation of those $p_i(t)$ is done exactly as in [14] starting from (E.20), then introducing the "charges" from (4.43) and the calculation of Appendix D for the influence functional and breaking up the tunnelling into small transition amplitudes into small timesteps dt yields the $\int_0^t \mathcal{D}\{t_{2n}\}$ included in the function $K_n^{(i)}(t)$ of Eqs. (4.62). The factor $K_0^{(i)}(t)$ is as in the one anharmonic bond case $+1$ of [14] by

4.2. TUNNELLING EXPECTATION VALUE USING EXTENDED NIBA 67

definition. The term $K_n^{(i)}(t)$, $i = 1, 2$ is defined as

$$K_n^{(i)}(t) = 2^{-(2n-1)} \sum_{\{\zeta_j^{(1)}, \zeta_{j'}^{(2)}\}} \sum_{\{\eta_j^{(1)}, \eta_{j'}^{(2)}\}} \int_0^t \mathcal{D}\{t_{2n}\} \mathcal{F}^n$$

$$\mathcal{F}^n = \mathcal{F}^n[\{t_j\}; \{\zeta_j^{(1)}\}; \{\eta_j^{(1)}\}; \{\zeta_{j'}^{(2)}\}; \{\eta_{j'}^{(2)}\}] \quad (4.70)$$

, where the summation of the blip- and sojourn-charges is explained later in Appendix F. The probability $p_i(t)$ is related to the expectation value $P_i(t)$ by the following relation

$$\langle \sigma_z^{(1)}(t) \otimes \sigma_z^{(2)}(t) \rangle \equiv P_i(t) = 2p_i(t) - 1, \; i = 1, 2 \quad (4.71)$$

$$P_i(t) = \sum_{n=0}^{\infty} (-1)^n \Delta^{2n} K_n^{(i)}(t), \quad i = 1, 2 \quad (4.72)$$

with

$$\int_0^t \mathcal{D}\{t_{2n}\} = \int_0^t dt_{2n} \int_0^{t_{2n}} dt_{2n-1} \cdots \int_0^{t_2} dt_1 \quad (4.73)$$

The functional $\mathcal{F}^n$ defined in Eq. (4.70) and derived in Appendix D can be split up into three parts

$$\mathcal{F}^n = \mathcal{F}_{(1)}^n[\{t_j\}; \{\zeta_j^{(1)}\}; \{\eta_j^{(1)}\}] \mathcal{F}_{(2)}^n[\{t_j\}; \{\zeta_{j'}^{(2)}\}; \{\eta_{j'}^{(2)}\}]$$
$$\cdot \mathcal{G}_{(12)}^n[\{t_j\}; \{\zeta_j^{(1)}\}; \{\eta_j^{(1)}\}; \{\zeta_{j'}^{(2)}\}; \{\eta_{j'}^{(2)}\}] \quad (4.74)$$

$$\mathcal{F}_{(1)}^n = \underbrace{e^{-\frac{q_0^2}{\pi\hbar} \sum_{j=1}^n Q_2^{(11)}(t_{2j}-t_{2j-1})}}_{\text{self-energy}} \underbrace{e^{-\frac{q_0^2}{\pi\hbar} \sum_{j'=1}^n \sum_{j=j'+1}^n \zeta_{j'}^{(1)} \zeta_j^{(1)} \Lambda_{jj'}^{(11)}}}_{\text{blip-blip interaction}} \underbrace{e^{\frac{iq_0^2}{\pi\hbar} \sum_{j'=0}^{n-1} \sum_{j=j'+1}^n \zeta_{j'}^{(1)} \eta_j^{(1)} X_{jj'}^{(11)}}}_{\text{blip-sojourn interaction}}$$

$$\mathcal{F}_{(2)}^n = \underbrace{e^{-\frac{q_0^2}{\pi\hbar} \sum_{j=1}^n Q_2^{(22)}(t_{2j}-t_{2j-1})}}_{\text{self-energy}} \underbrace{e^{-\frac{q_0^2}{\pi\hbar} \sum_{j'=1}^n \sum_{j=j'+1}^n \zeta_{j'}^{(2)} \zeta_j^{(2)} \Lambda_{jj'}^{(22)}}}_{\text{blip-blip interaction}} \underbrace{e^{\frac{iq_0^2}{\pi\hbar} \sum_{j'=0}^{n-1} \sum_{j=j'+1}^n \zeta_{j'}^{(2)} \eta_j^{(2)} X_{jj'}^{(22)}}}_{\text{blip-sojourn interaction}}$$

$$\mathcal{G}_{(12)}^n = \underbrace{e^{-\frac{q_0^2}{\pi\hbar} \sum_{j,j'=1}^n \zeta_j^{(1)} \zeta_{j'}^{(2)} \Lambda_{jj'}^{(12)}}}_{\text{blip-blip interaction}} \underbrace{e^{\frac{iq_0^2}{\pi\hbar} \left(\sum_{j'=0}^{n-1} \sum_{j=j'+1}^n \zeta_{j'}^{(1)} \eta_j^{(2)} X_{jj'}^{(12)} + \sum_{j=0}^{n-1} \sum_{j'=j+1}^n \zeta_{j'}^{(2)} \eta_j^{(1)} X_{j'j}^{(21)} \right)}}_{\text{blip-sojourn interaction}}$$

, where $\mathcal{F}_{(1)}^n$ describes the first anharmonic bond, $\mathcal{F}_{(2)}^n$ the second anharmonic bond and $\mathcal{G}_{(12)}^n$ the interaction between both bonds. The functionals $\mathcal{F}_{(1)}^n, \mathcal{F}_{(2)}^n$ can be treated exactly as in the one anharmonic bond case, since no interaction is present.

As in the one anharmonic bond case, the functions $Q_{1/2}(t)$ appear now, but with an additional index specifying the bond or the interaction between the two bonds.

$$\begin{aligned} Q_1^{(ab)}(t) &= \int_0^{\omega_0} d\omega \, \frac{J_{ab}(\omega)}{\omega^2} \sin(\omega t) \\ Q_2^{(ab)}(t) &= \int_0^{\omega_0} d\omega \, \frac{J_{ab}(\omega)}{\omega^2} (1 - \cos(\omega t)) \end{aligned} \quad (4.75)$$

The functions $\Lambda_{jj'}^{(ab)}, X_{jj'}^{(ab)}$ are defined as in [14], only extended for the case of two anharmonic bonds

$$\begin{aligned} \Lambda_{jj'}^{(ab)} &= Q_2^{(ab)}(t_{2j} - t_{2j'-1}) + Q_2^{(ab)}(t_{2j-1} - t_{2j'}) - Q_2^{(ab)}(t_{2j} - t_{2j'}) - Q_2^{(ab)}(t_{2j-1} - t_{2j'-1}) \\ X_{jj'}^{(ab)} &= Q_1^{(ab)}(t_{2j} - t_{2j'+1}) + Q_1^{(ab)}(t_{2j-1} - t_{2j'}) - Q_1^{(ab)}(t_{2j} - t_{2j'}) - Q_1^{(ab)}(t_{2j-1} - t_{2j'+1}) \end{aligned} \quad (4.76)$$

Applying the noninteracting blip approximation (NIBA), an approximation derived by [14] yields

$$\begin{aligned} \mathcal{F}_{(1),\text{NIBA}}^n &\equiv e^{-\frac{q_0^2}{\pi\hbar} \sum_{j=1}^n Q_2^{(11)}(t_{2j}-t_{2j-1})} \, e^{\frac{iq_0^2}{\pi\hbar} \sum_{j=1}^n \eta_{j-1}^{(1)} \varsigma_j^{(1)} Q_1^{(11)}(t_{2j}-t_{2j-1})} \\ \mathcal{F}_{(2),\text{NIBA}}^n &\equiv e^{-\frac{q_0^2}{\pi\hbar} \sum_{k=1}^m Q_2^{(22)}(t_{2j}-t_{2j-1})} \, e^{\frac{iq_0^2}{\pi\hbar} \sum_{j=1}^n \eta_{j-1}^{(2)} \varsigma_j^{(2)} Q_1^{(22)}(t_{2j}-t_{2j-1})} \end{aligned} \quad (4.77)$$

, where the same approximations as in the one anharmonic bond case (see [14] for details) have been performed

1. $X_{jj'}^{(aa)} = 0$, $j' \neq j - 1$, and put $X_{j,j-1}^{(aa)} = Q_1^{(aa)}(t_{2j} - t_{2j-1})$

2. $\Lambda_{jj'}^{(aa)} = 0$

But now the interaction part $\mathcal{G}_{(12)}^n$ requires an extension of the NIBA. The influence functional describing the interaction between both anharmonic bonds can be simplified by expanding the NIBA with the following requirements

$$\begin{aligned} \Lambda_{jj}^{(12)} &= 2Q_2^{(12)}(t_{2j} - t_{2j-1}) \\ \Lambda_{jj'}^{(12)} &= 0, \forall j \neq j' \\ X_{j,j-1}^{(12)} &= X_{j-1,j}^{(21)} = Q_1^{(12)}(t_{2j} - t_{2j-1}) \\ X_{j,j'}^{(12)} &= X_{j',j}^{(21)} = 0, \forall j \neq j' - 1 \end{aligned} \quad (4.78)$$

4.2. TUNNELLING EXPECTATION VALUE USING EXTENDED NIBA

, which results in the functional $\mathcal{G}^n_{(12),\text{NIBA}}$

$$\mathcal{G}^n_{(12),\text{NIBA}} = e^{-\frac{2q_0^2}{\pi\hbar}\sum_{j=1}^{n}\zeta_j^{(1)}\zeta_j^{(2)}Q_2^{(12)}(t_{2j}-t_{2j-1})} e^{\frac{iq_0^2}{\pi\hbar}\sum_{j=1}^{n}\left(\zeta_j^{(1)}\eta_{j-1}^{(2)}+\zeta_j^{(2)}\eta_{j-1}^{(1)}\right)Q_1^{(12)}(t_{2j}-t_{2j-1})}$$

The first approximation of the expanded NIBA lets only blips of different anharmonic bonds interact with each other at equal times. This keeps the interaction between both bonds alive and is also consistent with the normal NIBA. The interaction of blips of different anharmonic bonds at equal times, can be seen as a *self-energy* term between bond one and two. The second approximation lets the blip interact with its previous sojourn as in the one anharmonic bond case, but a blip of one anharmonic bond interacts with the sojourn of the other anharmonic bond, which preceded it in time.

$$P_i^{\text{NIBA}}(t) = \begin{cases} \sum_{n=0}^{\infty}(-1)^n\Delta^{2n}K_n^{(i),\text{NIBA}}(t), & i=1,2 \\ \sum_{n=1}^{\infty}(-1)^n\Delta^{2n}K_n^{(i),\text{NIBA}}(t), & i=3,4 \end{cases}$$

$$K_n^{(i),\text{NIBA}}(t) = 2^{-(2n-1)} \sum_{\{\zeta_j^{(1)},\zeta_j^{(2)}\}} \sum_{\{\eta_j^{(1)},\eta_j^{(2)}\}} \int_0^t \mathcal{D}\{t_{2n}\} \quad \mathcal{F}^n_{(1),\text{NIBA}}[\{t_j\};\{\zeta_j^{(1)}\};\{\eta_j^{(1)}\}]$$
$$\cdot \mathcal{F}^n_{(2),\text{NIBA}}[\{t_j\};\{\zeta_j^{(2)}\};\{\eta_j^{(2)}\}]$$
$$\cdot \mathcal{G}^n_{(12),\text{NIBA}}[\{t_j\};\{\zeta_j^{(1)}\};\{\eta_j^{(1)}\};\{\zeta_j^{(2)}\};\{\eta_j^{(2)}\}] \tag{4.79}$$

$$\mathcal{F}^n_{(1),\text{NIBA}} = \underbrace{e^{-\frac{q_0^2}{\pi\hbar}\sum_{j=1}^{n}Q_2(t_{2j}-t_{2j-1})}}_{\text{SE}_1} \underbrace{e^{\frac{iq_0^2}{\pi\hbar}\sum_{j=1}^{n}\eta_{j-1}^{(1)}\zeta_j^{(1)}Q_1(t_{2j}-t_{2j-1})}}_{\text{BS}_1}$$

$$\mathcal{F}^n_{(2),\text{NIBA}} = \underbrace{e^{-\frac{q_0^2}{\pi\hbar}\sum_{j=1}^{n}Q_2(t_{2j}-t_{2j-1})}}_{\text{SE}_2} \underbrace{e^{\frac{iq_0^2}{\pi\hbar}\sum_{j=1}^{n}\eta_{j-1}^{(2)}\zeta_j^{(2)}Q_1(t_{2j}-t_{2j-1})}}_{\text{BS}_2}$$

$$\mathcal{G}^n_{(12),\text{NIBA}} = \underbrace{e^{-\frac{2q_0^2}{\pi\hbar}\sum_{j=1}^{n}\zeta_j^{(1)}\zeta_j^{(2)}Q_2^{(12)}(t_{2j}-t_{2j-1})}}_{\text{BB}_{12}} \underbrace{e^{\frac{iq_0^2}{\pi\hbar}\sum_{j=1}^{n}\left(\zeta_j^{(1)}\eta_{j-1}^{(2)}+\zeta_j^{(2)}\eta_{j-1}^{(1)}\right)Q_1^{(12)}(t_{2j}-t_{2j-1})}}_{\text{BS}_{12}}$$

$$\tag{4.80}$$

Now the summation of the blip- and sojourn-charges has to be performed. For two anharmonic bonds this differs slightly, from the summation performed in [14].
A straightforward but tedious calculation[4] yields

$$K_n^{(i),\text{NIBA}}(t) = \int_0^t \mathcal{D}\{t_{2n}\} F_n^{(i)}(\{t_{2n}\})$$

---
[4]details presented in Appendix F

, with the functionals $F_n^{(i)}$, having been derived in Appendix F, of the following form

$$F_n^{(1)}(\{t_{2n}\}) = 2^{2n-1} \prod_{j=1}^n \cos\left(\frac{2q_0^2}{\pi\hbar} Q_1^+(t_{2j}-t_{2j-1})\right) \cdot e^{-\frac{2q_0^2}{\pi\hbar} Q_2^+(t_{2j}-t_{2j-1})}$$

$$F_n^{(2)}(\{t_{2n}\}) = 2^{2n-1} \prod_{j=1}^n \cos\left(\frac{2q_0^2}{\pi\hbar} Q_1^-(t_{2j}-t_{2j-1})\right) \cdot e^{-\frac{2q_0^2}{\pi\hbar} Q_2^-(t_{2j}-t_{2j-1})} \quad (4.81)$$

The factor of 2^{2n-1} in the term $F_n^{(i)}(\{t_{2n}\})$ cancels the factor of $2^{-(2n-1)}$ in $K_n^{(i),\,\mathrm{NIBA}}(t)$ of Eq. (4.79), as it should be, since this term covers the n blip-charge-pairs and $n-1$ sojourn-charge-pairs, because the initial and final sojourn-charges are fixed.

Defining the function $f_i(t)$, $i=1,2$ in the following form

$$f_1(t) = \Delta^2 \cos\left(\frac{2q_0^2}{\pi\hbar} Q_1^+(t)\right) \cdot e^{-\frac{2q_0^2}{\pi\hbar} Q_2^+(t)}$$

$$f_2(t) = \Delta^2 \cos\left(\frac{2q_0^2}{\pi\hbar} Q_1^-(t)\right) \cdot e^{-\frac{2q_0^2}{\pi\hbar} Q_2^-(t)} \quad (4.82)$$

Expressed in terms of $P_i^{\mathrm{NIBA}}(t)$

$$P_i^{\mathrm{NIBA}}(t) = \sum_{n=0}^\infty (-1)^n \int_0^t \mathcal{D}\{t_{2n}\} \prod_{j=1}^n f_i(t_{2j}-t_{2j-1}) \quad (4.83)$$

In order to calculate the $2n$-time integrations, a Laplace transform (see subsection 2.1.1 **Laplace Transform**) is helpful.

$$\tilde{P}_i^{\mathrm{NIBA}}(\lambda) = \int_0^\infty dt\, e^{-\lambda t} P_i^{\mathrm{NIBA}}(t)$$

$$= \sum_{n=0}^\infty (-1)^n \frac{\left(\tilde{f}_i(\lambda)\right)^n}{\lambda^{n+1}} = \frac{1}{\lambda + \tilde{f}_i(\lambda)} \quad (4.84)$$

, where $\tilde{f}_i(\lambda)$ is of course nothing but the Laplace transform of the earlier defined $f_i(t)$. Now inverting the Laplace transform, we are able to express $P_i^{\mathrm{NIBA}}(t)$ as

$$P_i^{\mathrm{NIBA}}(t) = \frac{1}{2\pi i} \int_{-i\infty+\delta}^{i\infty+\delta} d\lambda\, \frac{e^{\lambda t}}{\lambda + \tilde{f}_i(\lambda)} \quad (4.85)$$

Now the functions $\tilde{f}_i(\lambda)$, respectively $f_i(t)$ have to be calculated to perform the inverse Laplace transform. Since we are interested in the long time limit $\omega_0 t \gg 1$ one has to look first at the long time limit of the functions $Q_{1/2}^\pm(t)$. The following long time limits can easily be seen using

4.2. TUNNELLING EXPECTATION VALUE USING EXTENDED NIBA

the spectral densities $J_{ab}(\omega)$ of Eqs. (4.41), (4.42) and the definition Eq. (4.45).

$$
\begin{aligned}
Q_1^+(t) &\cong \frac{2C}{\omega_0}\left[\text{Si}(\omega_0 t) + \frac{(2-4D+D^2)(\omega_0 t \cos(\omega_0 t) - \sin(\omega_0 t))}{2(\omega_0 t)^2}\right] \stackrel{\omega_0 t \gg 1}{\approx} \frac{C\pi}{\omega_0} \\
Q_1^-(t) &\cong \frac{CD^2}{2\omega_0}\left[\frac{(\sin(\omega_0 t) - \omega_0 t \cos(\omega_0 t))}{(\omega_0 t)^2}\right] \stackrel{\omega_0 t \gg 1}{\approx} 0 \\
Q_2^+(t) &\cong \frac{C}{4\omega_0}\left[8\gamma - 8\text{Ci}(\omega_0 t) + 8\ln(\omega_0 t) - (2-4D+D^2)\right. \\
&\quad \left. + \frac{(4-8D+2D^2)(\cos(\omega_0 t) + \omega_0 t \sin(\omega_0 t) - 1)}{(\omega_0 t)^2}\right] \\
&\stackrel{\omega_0 t \gg 1}{\approx} \frac{C}{4\omega_0}\left(8\gamma + 8\ln(\omega_0 t) - (2-4D+D^2)\right) \stackrel{t \gg t_D}{\approx} \frac{2C}{\omega_0}\left(\gamma + \ln(\omega_0 t)\right) \\
Q_2^-(t) &\cong \frac{CD^2}{2\omega_0}\left[\frac{1-\cos(\omega_0 t) - \omega_0 t \sin(\omega_0 t)}{(\omega_0 t)^2} + \frac{1}{2}\right] \stackrel{\omega_0 t \gg 1}{\approx} \frac{CD^2}{4\omega_0}
\end{aligned} \quad (4.86)
$$

These long time limits are for the case of $Q_1^+(t), Q_2^-(t)$ constants, as in the phenomenological approach $Q_1^{\text{ohm}}(t), Q_2^{\text{superohm}}(t)$ used by Leggett et al. [14]. $Q_1^-(t), Q_2^+(t)$ behave in the long time limit like $Q_1^{\text{superohm}}(t), Q_2^{\text{ohm}}(t)$ in Leggett's phenomenological approach.

This strengthens the assumption of treating functions with the index "+" as a function showing ohmic dissipation, whereas the index "−" stands for super-ohmic dissipation. Leggett et al. do not have to consider the long time limit for the ohmic case, since they choose their spectral density phenomenologically and hence also their functions $Q_{1/2}(t)$. Their choice is made in a way, that allows further simplifications, but since we do not have this choice, the only possible way is an approximation regarding the physically interesting regime $\omega_0 t \gg 1$, which is achieved by the long time limit.

Next the long time approximation presented earlier, is applied for the Laplace transform of the functions $f_1(t), f_2(t)$ yielding

$$
\begin{aligned}
\tilde{f}_1(\lambda) &\cong \Delta^2 \omega_0^{-2\alpha} \cos(\alpha\pi) \Gamma(1-2\alpha) \lambda^{2\alpha-1} e^{-2\alpha\gamma} \\
\tilde{f}_2(\lambda) &\cong \Delta^2 \lambda^{-1} e^{-\frac{\alpha D^2}{4}}
\end{aligned} \quad (4.87)
$$

, where the parameter α plays the same role as in Leggett's article [14] and has the following definition

$$
\alpha = \frac{2q_0^2 C}{\pi \hbar \omega_0} \quad (4.88)
$$

The behaviour of the functions $\tilde{f}_1(\lambda), \tilde{f}_2(\lambda)$ presented above can be easily seen. The only relevant property for our case is the behaviour of those functions for the argument approaching zero and infinity. For $\lambda \to \infty$ all functions go to zero, as can easily be seen in Eqs. (4.87). The divergence for $\lambda \to 0$ may appear to give rise to complications. This is not the case. The function $\tilde{f}_1(\lambda)$ is of order $\lambda^{2\alpha-1}$, exactly as the function for the ohmic case in [14]. The function $\tilde{f}_2(\lambda)$ has a λ^{-1} pole, as the function describing super-ohmic behaviour in [14]. That again strengthens the assumption, that $P_i^{\text{NIBA}}(t)$ is either ohmic dissipative, for the case of $i=1$ and super-ohmic dissipative, for the case of $i=2$.

Now it is possible to treat the two different scenarios analogiously as in [14]. First we look at the poles of $P_1^{\text{NIBA}}(t)$.

$$\tilde{f}_1(\lambda) + \lambda = 0$$
$$\stackrel{\omega_0 t \gg 1}{\Longleftrightarrow} \Delta_{eff}^{2(1-\alpha)} \lambda^{2\alpha-1} e^{-2\alpha\gamma} + \lambda = 0 \tag{4.89}$$

, with the following definition for Δ_{eff}

$$\Delta_{eff} = [\Gamma(1-2\alpha)\cos(\pi\alpha)]^{\frac{1}{2(1-\alpha)}} \Delta \left(\frac{\Delta}{\omega_0}\right)^{\frac{\alpha}{1-\alpha}} \tag{4.90}$$

For $\alpha < \frac{1}{2}$ there are three poles. A branch-cut at $\lambda = 0$ and

$$\lambda_{p2/3} = \Delta_{eff}\, e^{-\frac{\alpha\gamma}{1-\alpha}}\, e^{\pm\frac{i\pi}{2(1-\alpha)}} \tag{4.91}$$

For $\frac{1}{2} < \alpha < 1$ the poles are not on the principal λ sheet and hence $P_1^{\text{NIBA}}(t)$ is given by the branch-cut. For $\alpha > 1$ the function $\tilde{f}(\lambda)$ no longer yields the leading factor in $\mathcal{O}(\lambda)$ with the factor $\lambda^{2\alpha-1}$. Now the term linear in λ is the leading term, hence we can write $\tilde{P}_1^{\text{NIBA}}(\lambda) \sim \lambda^{-1}$, yielding $P_1^{\text{NIBA}}(t) = 1$, which is nothing but the localisation phenomenon of Bray and Moore [26].

In the super-ohmic case we have to be careful. Up to now, we considered the NIBA, which requires considering the blips as a dilute gas. This is achieved by the self-energy of the blips, which reduces their "length" (in time) compared to the sojourn "length". The self-energy can be seen in the influence functional Eq. (4.74), by the term containing $Q_2(t)$. But in the super-ohmic case, this term approaches a constant for $t \to \infty$, whereas it reaches zero in the ohmic case (see Eqs. (4.86)). Due to this, the self-energy no longer suppresses the blip "length" compared to the sojourn "length", thus the blips cannot be considered as a dilute gas, and hence the NIBA appears not to be valid for the super-ohmic case.

The solution to this problem, has already been discussed in [14] and consists of a slight modification, which is explained below, that makes the NIBA still a valid approximation.

Splitting up the function $Q_2^-(t)$ into a constant and a time-dependent function $Q_3^-(t)$

$$\begin{aligned} Q_2^-(t) &\cong \frac{CD^2}{4\omega_0} + Q_3^-(t) \\ Q_3^-(t) &\cong \frac{CD^2}{2\omega_0}\left(\frac{1-\cos(\omega_0 t) - \omega_0 t \sin(\omega_0 t)}{(\omega_0 t)^2}\right) \end{aligned} \tag{4.92}$$

4.2. TUNNELLING EXPECTATION VALUE USING EXTENDED NIBA

As in [14] the time independent piece of $Q_2^-(t)$ will be absorbed into the level splitting Δ in the following way

$$\begin{aligned}\tilde{\Delta} &= \Delta e^{-\frac{2q_0^2}{\pi\hbar}Q_2^-(t=\infty)} \\ &= \Delta e^{-\frac{\alpha D^2}{4}}\end{aligned} \quad (4.93)$$

Now following [14] one can define a dimensionless quantity $b \ll 1$ in the following way

$$\frac{q_0^2}{\pi\hbar}Q_1^-(t=\tilde{\Delta}^{-1}), \frac{q_0^2}{\pi\hbar}Q_3^-(t=\tilde{\Delta}^{-1}) \sim \frac{q_0^2}{\hbar}\frac{J_-(\tilde{\Delta})}{\tilde{\Delta}} \equiv b \quad (4.94)$$

Pulling out the $\frac{1}{\lambda}$ pole, that can clearly be seen from the fact, that $Q_1^-(t)$ and $Q_3^-(t)$ approach zero for $t \to \infty$, one gets the following equation for finding the poles of $\tilde{P}(\lambda)$

$$\lambda^2 + \tilde{\Delta}^2\left(1 + \lambda\tilde{h}_-(\lambda)\right) = 0 \quad (4.95)$$

, with $\tilde{h}_-(\lambda)$ being the Laplace transform of $h_-(t)$ defined as

$$h_-(t) = \cos\left(\frac{2q_0^2}{\pi\hbar}Q_1^-(t)\right)e^{-\frac{2q_0^2}{\pi\hbar}Q_2^-(t)} - 1 \quad (4.96)$$

In the absence of damping ($h_- = 0$), the poles are entirely imaginary $\lambda = \pm i\tilde{\Delta}$. On physical grounds the poles will shift slightly off the imaginary axis and pick up a small negative real part. Expanding the above equation around the poles without damping, one finds

$$\lambda = \pm i\tilde{\Delta}\left(1 \pm \frac{i\tilde{\Delta}}{2} + ...\right) \quad (4.97)$$

Looking at the lowest order in b, the real and imaginary parts of λ are

$$\begin{aligned}\Im(\lambda) &= \pm\tilde{\Delta} \\ \Re(\lambda) &= -\frac{\tilde{\Delta}^2}{2}\Re(\tilde{h}_-(\lambda=i\tilde{\Delta})) \equiv -\Gamma_s \\ \Gamma_s &= \frac{\tilde{\Delta}^2}{2}\int_0^\infty dt\, \cos(\tilde{\Delta}t)h_-(t)\end{aligned} \quad (4.98)$$

Now expanding $h_-(t)$ in $Q_{1/3}^-(t)$ to lowest order of b yields

$$h_-(t) = \frac{2q_0^2}{\pi\hbar}Q_3^-(t) + \mathcal{O}(b^2) \quad (4.99)$$

With this expansion the integration can be performed

$$\begin{aligned}\Gamma_s &= \frac{q_0^2\tilde{\Delta}^2}{\pi\hbar}\int_0^\infty dt\, \cos(\tilde{\Delta}t)Q_3^-(t) \\ &= \frac{q_0^2\tilde{\Delta}^2}{\pi\hbar}\int_0^\infty dt\, \cos(\tilde{\Delta}t)\int_0^{\omega_0} d\omega\, \frac{J_-(\omega)}{\omega^2}\cos(\omega t)\end{aligned}$$

Now changing the order of integration, yields the final result for Γ_s

$$\begin{aligned}\Gamma_s &= \frac{q_0^2 \tilde{\Delta}^2}{\pi \hbar} \int_0^{\omega_0} d\omega \, \frac{J_-(\omega)}{\omega^2} \int_0^\infty dt \, \cos(\tilde{\Delta} t) \cos(\omega t) \\ &= \frac{q_0^2 \tilde{\Delta}^2}{2\pi \hbar} \int_0^{\omega_0} d\omega \, \frac{J_-(\omega)}{\omega^2} \int_{-\infty}^\infty dt \, \cos(\tilde{\Delta} t) \cos(\omega t) \\ &= \frac{q_0^2 \tilde{\Delta}^2}{2\hbar} \int_0^{\omega_0} d\omega \, \frac{J_-(\omega)}{\omega^2} \delta(\omega - \tilde{\Delta}) \\ &= \frac{q_0^2}{2\hbar} J_-(\tilde{\Delta}) \end{aligned} \qquad (4.100)$$

A self-consistent check of the smallness of the dimensionless quantity b, defined in Eq. (4.94), gives the following inequality

$$\frac{\Gamma_s}{\tilde{\Delta}} \ll 1$$

Using the results and definitions of Eqs. (4.93) and (4.92) one obtains

$$\begin{aligned}\frac{\Gamma_s}{\tilde{\Delta}} &\cong \frac{J_-(\tilde{\Delta})}{\tilde{\Delta}} \\ &= \frac{CD^2}{2\omega_0} \left(\frac{\Delta}{\omega_0}\right)^2 e^{-\frac{\alpha D^2}{2}} \end{aligned} \qquad (4.101)$$

We know, that $\frac{\Delta}{\omega_0}$ is much smaller than one. As D^2 increases quadratically, the exponential factor also containing D^2, leads to fulfilment of the above inequality.

Performing an inverse Laplace transform of $\tilde{P}_2^{\text{NIBA}}(\lambda)$, with the above calculated complex conjugate poles $\lambda = -\Gamma_s \pm i\tilde{\Delta}$, gives the final result

$$P_2^{\text{NIBA}}(t) = \cos(\tilde{\Delta} t) e^{-\Gamma_s t} \qquad (4.102)$$

This describes under-damped coherent oscillations at frequency $\tilde{\Delta}$ with the damping-rate Γ_s for the case of super-ohmic dissipation described by the function $f_1(t)$.

4.2.1 Summary

First of all one has to look at how $P(t)$ behaves depending on different initial positions of the anharmonic bonds. What we see, is that the overall tunnelling process, by which starting from an initial configuration and reaching a final configuration is meant, does not only depend on the initial position, but on the relation of the initial configuration to the final configuration.

4.2. TUNNELLING EXPECTATION VALUE USING EXTENDED NIBA

The four tunnelling probabilities hence can be reduced to two different scenarios, because of the simple relation of Eq. (4.63).

The two different overall tunnelling processes describe either tunnelling with length-change, which is described by $P_1(t)$ and shows ohmic dissipation and tunnelling without length-change, which results in super-ohmic dissipation. For $P_1(t)$ a phase transition occurs depending on α, since now we have purely ohmic dissipation in every tunnelling transition. For $\alpha < 1$ there is no localisation, that means both anharmonic bonds spend on average the same time in each of the two equilibrium positions, whereas the symmetry becomes broken for $\alpha \geq 1$ leading to localisation. That means both anharmonic bonds spend on average most of their time in the equilibrium position they were initially prepared in. This has already been calculated for the purely phenomenological choice of the spectral density of [14]. The phase transition has been found by [26], as mentioned before. $P_2(t)$ exhibits super-ohmic dissipation and hence tunnelling is never surpressed.

Chapter 5

Results and Conclusions

We considered a microscopic model system in 1 dimension with N particles. All particles interact with their nearest neighbour described by a harmonic potential except the r anharmonic bonds. Those anharmonic bonds interact through a symmetric double well potential with each other. The coupling between the harmonic and the anharmonic bonds have been described by the coupling constants c_σ. At first we looked at the simplest case of one anharmonic bond ($r = 1$). Two methods were presented to analytically separate the harmonic form the anharmonic degrees of freedom. The first method is intuitive, but only applicable to one dimension, whereas the second can be generalized to d-dimensions. A third method, which is not presented in this thesis, is shown in the publication [36]. This analytical separation of the harmonic from the anharmonic degrees of freedom allows us to derive the up to now phenomenologically considered Caldeira-Leggett Hamiltonian, analytically from a microscopic model.

Next we consider the position dependence of the anharmonic bond in the tunnelling behaviour. As a result we get, that if the anharmonic bond is located at the border of the chain, we have a transition from ohmic to super-ohmic dissipation, which is seen in the frequency dependence of c_σ exhibiting a sensitivity of the location of the anharmonic bond M. That means, the anharmonic bond tunnels between the two minima of the potential, dissipating energy to the harmonic bath around it. The terms ohmic and super-ohmic refer to the way the energy is dissipated. The dissipation for low frequencies has the form of a power-law $\sim \omega^s$, where the term ohmic stands for $s = 1$, while super-ohmic refers to $s > 1$. The super-ohmic terms calculated in this thesis yielded $s = 3$ for the considered model.

As shown in [26] the quantum-mechanical tunnelling in a symmetric double well potential under the influence of a harmonic bath introducing dissipation, can be mapped one the one-dimensional Ising-model with inverse-square-law interactions R^{-2} for the ohmic case and interactions falling off like R^{-4} in the super-ohmic case. As has been shown by Thouless [37] an interaction energy falling off like R^{-n} in an one-dimensional system, shows a phase transition from an "ordered" to a "disordered" phase at $n = 2$.

The expression "ordered" can be interpreted as the anharmonic bond spending most of its time in the well it has been initially prepared in. The case of "disordered" refers to the anharmonic bond spending half of its time in one well and the the other half in the other well.

From this we can see, that the case of super-ohmic dissipation refers to the disordered phase and hence tunnelling is never suppressed. If the anharmonic bond is located in the bulk, the system shows a much more interesting behaviour. Calculations show, that in this case the system shows only ohmic dissipation. Ohmic dissipation $\sim \omega$ described in language of the Ising-model, has an interaction falling off as R^{-2}. This is exactly the critical exponent [37] in an infinite one-dimensional system, where a phase transition occurs. For a coupling constant below a critical value, the system will show ohmic dissipative behaviour, where the anharmonic bond tunnels between the two wells as in the super-ohmic case. But for a coupling stronger than the critical coupling constant, the system will stop tunnelling back and forth and will remain most of its time in the well it started in. This is a spontaneous symmetry breaking, which is of the same universality class as the one-dimensional Ising model with inverse-square-law interactions [26].

The next step was to include a second anharmonic bond into the one-dimensional model. This leads to indirect interaction of both anharmonic bonds through the environment. There are many possible positions of both bonds, but we focus on both anharmonic bonds located in the bulk with a finite, but variable distance D between them.

The choice of a finite distance D can be explained as follows. For the case of *infinite* distance, both bonds do no longer interact indirectly with each other and hence the system reduces to two isolated anharmonic bonds and their position dependence. This has already been discussed in the first part of this thesis. The choice of both anharmonic bonds located in the bulk is due to the following argument. As we saw in the first part, the position of the anharmonic bond in the bulk showed ohmic dissipation and a phase transition for a coupling constant higher than the critical value. Since we clearly want to show, that the transition is due to the indirect interaction and **not** due to both bonds being at one of the borders, we chose to consider only the case of both bonds being in the bulk.

The first problem that had to be solved was the analytical diagonalisation procedure. Following the first method presented in chapter 3.1, we had to choose the anharmonic bond positions M_1, M_2 symmetrically around the center of the chain. This allowed to replace the bond positions and the total chain-length N by just two parameters D, N.

Next the calculation of the kernel followed. The kernel appearing in the influence Euclidean action is coupled to both anharmonic bonds. That is the reason a mapping like in the one anharmonic bond case, as done by Bray and Moore [26], is no longer possible. With a transformation the coupling can be eliminated in the influence part, but it is only shifted to the local part. This results in an direct interaction of the instantons of both anharmonic bonds, whose effect is not fully understood. The kernel $K_D^{++}(\tau)$, which is achieved after the transformation in the influence action exhibits ohmic dissipative behaviour, whereas the the kernel $K_D^{--}(\tau)$ shows a transition to super-ohmic dissipative behaviour for $\tau \gg \tau_D$. Since the coupling of the instantons in the local part of the influence action occurs, we wanted to put these results on more stable ground and hence chose to calculate the tunnelling probability in the way Leggett et al. [14] did, but generalized to two anharmonic bonds interacting indirectly through the environment.

Using the Feynman-Vernon technique [21] to eliminate the harmonic degrees of freedom via the path integral formalism, we were able to express the tunnelling probabilities. A necessary restriction was to set the tunnelling times of both anharmonic bonds equal. This restriction has also been applied by Dubé and Stamp [29], but without mathematical rigour and with incomprehensible argumentations. We already imply this restrictions at an earlier point, namely when defining the bare tunnelling amplitudes using the spin-boson approach applied by Leggett et al. [14]. This of course simplifies the calculations of the tunnelling probability $p(t)$, but also allows only two physically different scenarios labelled as $p_1(t)$ and $p_2(t)$. The tunnelling probability $p_1(t)$ stands for an initial and final position, where both anharmonic bonds are in equal equilibrium positions a_s or a_l. The tunnelling is split up in $2n$ tunnelling transitions, where each transitions means both anharmonic bonds changing their equilibrium positions from a_s to a_l or the other way around. The calculated spectral density results in $p_1(t)$ showing only ohmic dissipation. This is understandable, since ohmic dissipation already appeared for one anharmonic bond located in the bulk in chapter 3.3. Tunnelling requires as in the one anharmonic bond case a movement of an infinite mass, which leads to ohmic dissipative tunnelling. The other scenario $p_2(t)$ implies both anharmonic bonds having different initial equilibrium length a_s, a_l and final positions. Because of the restriction of both anharmonic bonds tunnelling at equal times, one of the bonds tunnels from a_s to a_l, whereas the other bond does just the opposite. The total length of the anharmonic bonds stays the same at each of the $2n$ tunnelling transitions. The calculated spectral density for this case is purely super-ohmic. For the case of one anharmonic bond located at the border of the chain in chapter 3.3, the spectral density is also super-ohmic, hence the result achieved for $p_2(t)$ is not surprising. The calculation of the function $p_i(t)$, $i = 1, 2$ is done in analogy to [14] using the NIBA and extending it to the case of two anharmonic bonds. The extension of the NIBA allows coupling of blips of both anharmonic bonds at the same time, which can be interpreted as a blip-self-energy. The other extension is a blip not only coupling with its previous sojourn, but also with the previous sojourn of the other anharmonic bond.

The final results for the tunnelling probability are achieved as in the one anharmonic case solved by Leggett et al. [14], by Laplace transformation and investigation of the poles.

Appendix A

Diagonalisation of the first approach

The separation of the harmonic and anharmonic degrees of freedom by using centre of mass and relative coordinates of the *total* chain has been described in section "First Method". The transformation to normal coordinates requires the diagonalisation of (T_{kj}). In the present Appendix the steps of the diagonalisation procedure will be given. Making use of Eqs. (3.14), (3.15), (3.17) - (3.19) one obtains the harmonic part of the Hamiltonian Eq. (3.12) with the following symmetric matrix (T_{kj})

$$T_{ii} = \begin{cases} 2, & i = 1, ..., M-2, M+2, ..., N-1 \\ \frac{3}{2}, & i = M-1, M+1 \end{cases}$$

$$T_{i,i+1} = \begin{cases} -1, & i = 1, ..., M-2, M+1, ..., N-2 \\ 0, & i = M-1 \end{cases}$$

$$T_{i,i+2} = \begin{cases} 0, & i = 1, ..., M-3, M+1, ..., N-3 \\ -\frac{1}{2}, & i = M-1 \end{cases} \quad (A.1)$$

Then it is straightforward to solve the eigenvalue equation

$$\sum_{\substack{j=1 \\ (j \neq M)}}^{N-1} T_{kj} u_j^{(\sigma)} = \lambda'_\sigma u_k^{(\sigma)}, \quad \sigma = 1, ..., N-2 \quad (A.2)$$

Writing this equation explicitly, gives

$$(2 - \lambda'_\sigma) u_1^{(\sigma)} - u_2^{(\sigma)} = 0, \quad k = 1 \tag{A.3}$$

$$(2 - \lambda'_\sigma) u_k^{(\sigma)} - \left(u_{k+1}^{(\sigma)} + u_{k-1}^{(\sigma)}\right) = 0, \quad k = 2, ..., M-2, M+2, ..., N-2 \tag{A.4}$$

$$\left(\frac{3}{2} - \lambda'_\sigma\right) u_{M-1}^{(\sigma)} - u_{M-2}^{(\sigma)} - \frac{1}{2} u_{M+1}^{(\sigma)} = 0, \quad k = M-1 \tag{A.5}$$

$$\left(\frac{3}{2} - \lambda'_\sigma\right) u_{M+1}^{(\sigma)} - u_{M+2}^{(\sigma)} - \frac{1}{2} u_{M-1}^{(\sigma)} = 0, \quad k = M+1 \tag{A.6}$$

$$(2 - \lambda'_\sigma) u_{N-1}^{(\sigma)} - u_{N-2}^{(\sigma)} = 0, \quad k = N-1 \tag{A.7}$$

The form of Eq. (A.4) suggests an ansatz of plain waves for the left and the right part of the form

$$u_k^{(\sigma)} = \begin{cases} A_+ e^{iq_\sigma k} + A_- e^{-iq_\sigma k} & k = 1, ..., M-2 \\ B_+ e^{iq_\sigma k} + B_- e^{-iq_\sigma k} & k = M+2, ..., N-1 \end{cases} \tag{A.8}$$

Using this ansatz it is possible to include Eq. (A.3) in Eq. (A.4) and Eq. (A.7) in Eq. (A.6) using the requirements $u_0^{(\sigma)} = u_N^{(\sigma)} \stackrel{!}{=} 0$. These requirements are boundary conditions of an **open** chain, that lead to $q_\sigma \in (0, \pi)$. 0 and π are not included, because these values do not lead to non-zero eigenvectors. Applying these boundary conditions yields

$$\begin{aligned} u_0^{(\sigma)} &= A_+ + A_- \stackrel{!}{=} 0 & \Leftrightarrow A_- = -A_+ \\ u_N^{(\sigma)} &= B_+ e^{iq_\sigma N} + B_- e^{-iq_\sigma N} \stackrel{!}{=} 0 & \Leftrightarrow B_- = -B_+ e^{2iq_\sigma N} \end{aligned} \tag{A.9}$$

Using those results in Eqs. (A.4) and (A.6) leads to

$$u_k^{(\sigma)} = \mathcal{N}_{b_\sigma} \begin{cases} \sin(q_\sigma k) & , \quad k = 1, ..., M-2; \quad \mathcal{N}_{b_\sigma} = 2iA_+ \\ b_\sigma \sin(q_\sigma [N-k]) & , \quad k = M+2, ..., N-1; \quad b_\sigma = \frac{B_+}{A_+} e^{iq_\sigma N} \end{cases}$$

$$\lambda'_\sigma = 2(1 - \cos(q_\sigma)) = m\lambda_\sigma \tag{A.10}$$

The eigenvector components $u_{M-1}^{(\sigma)}, u_{M+1}^{(\sigma)}$ can be calculated straightforward by using the results obtained and plugging them into Eq. (A.4) for the cases of $k = M \pm 2$. Hence the full set of eigenvectors reads

$$u_k^{(\sigma)} = \mathcal{N}_{b_\sigma} \begin{cases} \sin(q_\sigma k), & 1 \leq k \leq M-1 \\ b_\sigma \sin(q_\sigma [N-k]), & M+1 \leq k \leq N-1 \end{cases} \tag{A.11}$$

with $\mathcal{N}_\sigma$ as the normalisation constant and b_σ being a coefficient depending on the wave number q_σ, the location M of the anharmonic bond and the total length of the chain N. The coefficient b_σ can be obtained from Eq. (A.5) by straightforward calculation, yielding

$$b_\sigma = \frac{2\sin(q_\sigma M) - \sin(q_\sigma [M-1])}{\sin(q_\sigma [N-M-1])} \tag{A.12}$$

Using (A.6) and the result obtained from (A.12), it is possible to obtain a transcendental equation of the form

$$\sin(q_\sigma[N-M-1])\sin(q_\sigma[M-1]) - \Big(2\sin(q_\sigma[N-M]) - \sin(q_\sigma[N-M-1])\Big)$$
$$\cdot \Big(2\sin(q_\sigma M) - \sin(q_\sigma[M-1])\Big) = 0 \quad . \quad (A.13)$$

This equation cannot be solved analytically. Hence a separation of the parameters N, M using trigonometric identities is useful for a further discussion. The main steps of the separation will be shown. By performing the multiplication of the two factors the transcendental equation can be put into

$$\sin(q_\sigma[N-M-1])\sin(q_\sigma M) + \sin(q_\sigma[N-M])\sin(q_\sigma[M-1])$$
$$-2\sin(q_\sigma[N-M])\sin(q_\sigma M) = 0 \quad (A.14)$$

now separating the parameters N, M and assuming $\sin(q_\sigma N) \neq 0$ leads to

$$\sin(q_\sigma N)\Big[\Big(\cos(q_\sigma[M+1])\sin(q_\sigma M) + \cos(q_\sigma M)\sin(q_\sigma[M-1]) - 2\cos(q_\sigma M)\sin(q_\sigma M)\Big)$$
$$-\cot(q_\sigma N)\Big(\sin(q_\sigma[M+1])\sin(q_\sigma M) + \sin(q_\sigma M)\sin(q_\sigma[M-1]) - 2\sin^2(q_\sigma M)\Big)\Big] = 0$$

Using the assumption $\sin(q_\sigma M) \neq 0$ allows to write the equation as

$$\cot(q_\sigma N) = \frac{\cos(q_\sigma[M+1]) + \cot(q_\sigma M)\sin(q_\sigma[M-1]) - 2\cos(q_\sigma M)}{\sin(q_\sigma[M-1]) + \sin(q_\sigma[M+1]) - 2\sin(q_\sigma M)}$$

The identity $\sin(x) + \sin(y) = 2\sin\left(\frac{x+y}{2}\right)\cos\left(\frac{x-y}{2}\right)$ applied to the denominator gives

$$\cot(q_\sigma N) = \frac{2\cos(q_\sigma M) - \cos(q_\sigma[M+1]) - \cot(q_\sigma M)\sin(q_\sigma[M-1])}{4\sin(q_\sigma M)\sin^2\left(\frac{q_\sigma}{2}\right)}$$

In the numerator the parameter M can be isolated leading to

$$\cot(q_\sigma N) = \frac{2\cos(q_\sigma M)\Big(1 - \cos(q_\sigma)\Big) + \sin(q_\sigma M)\sin(q_\sigma)\Big(1 + \cot^2(q_\sigma M)\Big)}{4\sin(q_\sigma M)\sin^2\left(\frac{q_\sigma}{2}\right)}$$

Now the two additive factors of the numerator can be separated and using basic trigonometric identities one gets

$$\cot(q_\sigma N) = \underbrace{\cot(q_\sigma M) + \frac{\cot\left(\frac{q_\sigma}{2}\right)}{2\sin^2(q_\sigma M)}}_{f(q_\sigma)} \quad (A.15)$$

Since the transcendental Eq. (A.15) is not analytically solvable a detailed discussion for approximative solutions is given. The l.h.s. behaves as follows

$$\lim_{q_\sigma \searrow \frac{\pi\sigma}{N}} \cot(q_\sigma N) = \infty, \qquad \lim_{q_\sigma \nearrow \frac{\pi\sigma}{N}} \cot(q_\sigma N) = -\infty, \qquad \sigma = 0, ..., N \qquad (A.16)$$

which means the l.h.s. *oscillates* in every interval $\left[\frac{\pi\sigma}{N}, \frac{\pi(\sigma+1)}{N}\right]$, $\sigma = 0, ..., N-1$ from ∞ to $-\infty$. The r.h.s. shows this behaviour:

$$\lim_{q_\sigma \searrow \frac{\pi\sigma}{M}} f(q_\sigma) = \infty, \qquad \lim_{q_\sigma \nearrow \frac{\pi\sigma}{M}} f(q_\sigma) = \infty, \qquad \sigma = 1, ..., M-1$$

$$\lim_{q_\sigma \nearrow \pi} f(q_\sigma) = -\infty \qquad (A.17)$$

Since $M < N$ and with the results obtained from Eqs. (A.16), (A.17), it is easy to see that there is exactly one solution in every interval $\left[\frac{\pi\sigma}{N}, \frac{\pi(\sigma+1)}{N}\right]$ for $\sigma = 1, ..., N-2$, leading to the expected number of $N-2$ solutions. The remaining two degrees of freedom are the centre of mass and the anharmonic bond coordinate X_c and q_M, respectively. The solutions can hence be written as

$$q_\sigma = \frac{\pi}{N} \cdot \sigma + \epsilon_\sigma, \qquad \sigma = 1, ..., N-2 \qquad (A.18)$$

with $0 \leq \epsilon_\sigma < \frac{\pi}{N}$. That means ϵ_σ is of the form $\epsilon_\sigma = N^{-\alpha}$, $\alpha \geq 1$. For $\sigma = \mathcal{O}(1)$ the form of ϵ_σ can be determined from the transcendental equation by applying Eq. (A.18) and using basic trigonometric identities and the following basic approximations

$$\cos(x \pm y) = \cos(x)\cos(y) \mp \sin(x)\sin(y)$$
$$\sin(x \pm y) = \sin(x)\cos(y) \pm \cos(x)\sin(y)$$
$$\sin(x) \stackrel{x \ll 1}{\approx} x \qquad (A.19)$$
$$\cos(x) \stackrel{x \ll 1}{\approx} 1 \qquad (A.20)$$

The transcendental equation can be written as

$$\frac{\cos(\epsilon_\sigma N)}{\sin(\epsilon_\sigma N)} = \underbrace{\frac{\cos\left(\frac{\pi\sigma M}{N}\right)\cos(\epsilon_\sigma M) - \sin\left(\frac{\pi\sigma M}{N}\right)\sin(\epsilon_\sigma M)}{\cos\left(\frac{\pi\sigma M}{N}\right)\sin(\epsilon_\sigma M) + \sin\left(\frac{\pi\sigma M}{N}\right)\cos(\epsilon_\sigma M)}}_{\mathcal{O}(1)} + \underbrace{\frac{1 - \frac{\pi\sigma}{4N}\epsilon_\sigma}{2\left[\frac{\pi\sigma}{2N} + \frac{\epsilon_\sigma}{2}\right]\left[\sin\left(\frac{\pi\sigma M}{N}\right) + \frac{\epsilon_\sigma M}{2}\cos\left(\frac{\pi\sigma M}{N}\right)\right]^2}}_{\mathcal{O}(N)} \qquad (A.21)$$

where it is obvious that the first term is of $\mathcal{O}(1)$, since the denominator is non-zero and all trigonometric functions are of $\mathcal{O}(1)$. The second term shows a numerator of $\mathcal{O}(1)$ and a denominator of $\mathcal{O}\left(\frac{1}{N}\right)$ in leading order. To fulfil this equation the l.h.s. **must** be of order $\mathcal{O}(N)$.

This can only be achieved if $\epsilon_\sigma = \mathcal{O}(N^{-\alpha})$, $\alpha > 1$. By assuming this and Taylor expanding the l.h.s. one sees immediately that ϵ_σ is of order $\mathcal{O}\left(\frac{1}{N^2}\right)$. The exact result is

$$\epsilon_\sigma = \frac{\pi\sigma \sin^2\left(\frac{\pi\sigma M}{N}\right)}{N^2} + \mathcal{O}\left(\frac{1}{N^3}\right) \tag{A.22}$$

In the thermodynamic limit $N \to \infty$, q_σ becomes continuous within $(0, \pi)$ with constant density which implies a constant low energy density of states.

The normalisation constant $\mathcal{N}_\sigma$ and the coefficient b_σ are functions of q_σ, M and N. The normalisation reads explicitly

$$\begin{aligned}
1 &= \mathcal{N}_{b_\sigma}^2 \left[\sum_{k=1}^{M-1} \sin^2(q_\sigma k) + b_\sigma^2 \sum_{k=M+1}^{N-1} \sin^2(q_\sigma[N-k]) \right] \\
&= \frac{\mathcal{N}_{b_\sigma}^2}{2} \left[\sum_{k=1}^{M-1} \left(1 - \cos(2q_\sigma k)\right) + b_\sigma^2 \sum_{k=M+1}^{N-1} \left(1 - \cos(2q_\sigma[N-k])\right) \right] \\
&= \frac{\mathcal{N}_{b_\sigma}^2}{2} \left[(M-1) + b_\sigma^2(N-M-1) - \sum_{k=1}^{M-1} \cos(2q_\sigma k) - b_\sigma^2 \sum_{k'=1}^{N-M-1} \cos(2q_\sigma k') \right]
\end{aligned}$$

$$\Rightarrow \mathcal{N}_{b_\sigma} = \sqrt{\frac{2}{(M-1) + b_\sigma^2(N-M-1) - \frac{\sin(q_\sigma[M-1])\cos(q_\sigma M) + b_\sigma^2 \sin(q_\sigma(N-M-1))\cos(q_\sigma[N-M])}{\sin(q_\sigma)}}} \tag{A.23}$$

and in the limit of large N (or low frequency behaviour $\omega_\sigma \ll 1$) we get

$$\mathcal{N}_{b_\sigma} \sim \sqrt{\frac{2}{N}} \tag{A.24}$$

Appendix B

Diagonalisation of the second approach

In this Appendix the diagonalisation procedure of the harmonic coordinates will be discussed, since in the second approach the momenta already are diagonal. As stated in section "Second Method" equal masses $m_n = m$ are considered. Applying Eq. (3.35) to Eq. (3.34) yields

$$
\begin{aligned}
H &= H_d + H_{harm} + H_{int} \\
H_d &= \frac{\pi_M^2}{m} + V_0(q_M) + \frac{C}{4}q_M^2 \\
H_{harm} &= \sum_{\substack{n=1 \\ n \neq M, M+1}}^{N-1} \frac{p_n'^2}{2m} + \underbrace{\frac{C}{2}\sum_{n=1}^{N-2}(x_{n+1}' - x_n' - a_n')^2}_{V_{harm}(\{x_n'\})} \\
H_{int} &= -\frac{C}{2}\left(x_{M+1}' - x_{M-1}' - a_M' - a_{M-1}'\right)q_M
\end{aligned}
\qquad (B.1)
$$

A transformation

$$
\begin{aligned}
x_n'^{(eq)} &= x_1' + \sum_{i=1}^{n-1} a_i' \\
x_n' &= x_n'^{(eq)} + u_n'
\end{aligned}
\qquad (B.2)
$$

defining an equilibrium position $x_n'^{(eq)}$, allows to rewrite the harmonic Hamiltonian as

$$
H_{harm} = \sum_{\substack{n=1 \\ n \neq M, M+1}}^{N-1} \frac{p_n'^2}{2m} + \frac{C}{2}\sum_{n=1}^{N-2}(u_{n+1}' - u_n')^2 \qquad (B.3)
$$

Expanding the harmonic potential $V_{harm}(\{x_n'\})$ around the equilibrium configuration $V_{harm}\left(\{x_n'^{(eq)}\}\right)$ up to quadratic order, yields

$$
V_{harm}(x_n') = \underbrace{V\left(x_n'^{(eq)} + u_n'\right)}_{\text{constant}} + \underbrace{\frac{\partial V_{harm}}{\partial x_k'}\left(x_n'^{(eq)}\right)u_k'}_{0} + \frac{1}{2}\underbrace{\frac{\partial^2 V_{harm}}{\partial x_k' \partial x_l'}\left(x_n'^{(eq)}\right)}_{V_{kl}'}u_k' u_l' \qquad (B.4)
$$

APPENDIX B. DIAGONALISATION OF THE SECOND APPROACH

This is **not** an approximation, since V_{harm} is a harmonic potential. The variable u'_n represents the displacement of x'_n from the equilibrium configuration $x'^{(eq)}_n$.

Introducing mass weighted coordinates

$$\begin{aligned} \tilde{u}'_n &= \sqrt{m}\, u'_n \\ \tilde{p}'_n &= \frac{p'_n}{\sqrt{m}} \end{aligned} \qquad (B.5)$$

yields the harmonic Hamiltonian

$$H_{harm} = \frac{1}{2}\sum_{n=1}^{N-1}(\tilde{p}'_n)^2 + \frac{1}{2}\sum_{k,l=1}^{N-1}\tilde{V}'_{kl}\tilde{u}'_k\tilde{u}'_l \qquad (B.6)$$

With the nonzero elements of the symmetric matrix $\left(\tilde{V}'_{kl}\right)$ read explicitly

$$\tilde{V}'_{kk} = \frac{C}{m}\begin{cases} 1, & k=1, M, N-1 \\ 2, & k=2,...,N-2 \end{cases}$$

$$\tilde{V}'_{k,k+1} = \frac{C}{m}\begin{cases} -1, & k=1, M-2, M+1 N-2 \\ -\frac{1}{\sqrt{2}}, & k=M-1, M \end{cases} \qquad (B.7)$$

Diagonalising this matrix in the standard way

$$\sum_{\substack{k=1 \\ (\neq M\pm 1, M)}}^{N-1} \tilde{V}'_{kl} e^{(\sigma)}_k = \tilde{\lambda}_\sigma\, e^{(\sigma)}_l \qquad (B.8)$$

and considering the remaining equations ($k = M-1, M, M+1$), which yield a non-trivial solution if a corresponding determinant vanishes. This leads to the following mass weighted eigenvectors and eigenvalues (the calculation for $k \neq M-1, M, M+1$ is absolutely analogous to the procedure in Appendix A Eqs. (A.3)-(A.11)) and the determinant condition is a straightforward calculation, yielding

$$\begin{aligned} e^{(\sigma)}_n &= \tilde{\mathcal{N}}_{\tilde{b}_\sigma}\begin{cases} \cos\left(\tilde{x}_\sigma[n-\tfrac{1}{2}]\right), & n=1,...,M-1 \\ \tilde{b}_\sigma \cos\left(\tilde{x}_\sigma[N-n-\tfrac{1}{2}]\right), & n=M+1,...,N-1 \end{cases} \\ e^{(\sigma)}_M &= \sqrt{2}\tilde{\mathcal{N}}_{\tilde{b}_\sigma}\cos\left(\tilde{x}_\sigma\left[M-\frac{1}{2}\right]\right) \\ \tilde{\lambda}_\sigma &= \frac{2C}{m}(1-\cos(\tilde{x}_\sigma)) \end{aligned} \qquad (B.9)$$

where $\tilde{x}_\sigma$ are the wave numbers used in the ansatz (which is equivalent in the form to the ansatz used in Appendix A Eq. (A.8)). The other parameters are the normalisation constant $\tilde{\mathcal{N}}_{\tilde{b}_\sigma}$ and a coefficient $\tilde{b}_\sigma$ describing the position dependence of the anharmonic bond. The parameters are functions depending on $\tilde{x}_\sigma, N$ and M. Their explicit expression are

$$\tilde{b}_\sigma = \frac{\cos\left(\tilde{x}_\sigma\left[M - \frac{1}{2}\right]\right)}{\cos\left(\tilde{x}_\sigma\left[N - M - \frac{1}{2}\right]\right)} \tag{B.10}$$

the calculation of the normalisation constant

$$\begin{aligned}
1 &= \tilde{N}_{\tilde{b}_\sigma}^2 \left[\sum_{k=1}^{M-1} \cos^2\left(\tilde{x}_\sigma\left[k - \frac{1}{2}\right]\right) + 2\cos^2\left(\tilde{x}_\sigma\left[M - \frac{1}{2}\right]\right)\right. \\
&\quad \left. + \tilde{b}_\sigma^2 \sum_{k=M+1}^{N-1} \cos^2\left(\tilde{x}_\sigma\left[N - k - \frac{1}{2}\right]\right)\right] \\
&= \frac{\tilde{N}_{\tilde{b}_\sigma}^2}{2}\left[\sum_{k=1}^{M}\left(1 + \cos(2\tilde{x}_\sigma k)\cos(\tilde{x}_\sigma) + \sin(2\tilde{x}_\sigma k)\sin(\tilde{x}_\sigma)\right) + 2\cos^2\left(\tilde{x}_\sigma\left[M - \frac{1}{2}\right]\right)\right. \\
&\quad \left. + \tilde{b}_\sigma^2 \sum_{k=M+1}^{N-1}\left(1 + \cos(2\tilde{x}_\sigma[N-k])\cos(\tilde{x}_\sigma) + \sin(2\tilde{x}_\sigma[N-k])\sin(\tilde{x}_\sigma)\right)\right]
\end{aligned} \tag{B.11}$$

yields

$$\begin{aligned}
\tilde{N}_{\tilde{b}_\sigma} &= \sqrt{\frac{2}{\text{denom}}} \\
\text{denom} &= M - 1 + \tilde{b}_\sigma^2[N - M - 1] + 2\cos^2\left(\tilde{x}_\sigma\left[M - \frac{1}{2}\right]\right) \\
&\quad + \frac{\sin(\tilde{x}_\sigma[M-1])\cos(\tilde{x}_\sigma[M-1]) + \tilde{b}_\sigma^2\sin(\tilde{x}_\sigma[N-M-1])\cos(\tilde{x}_\sigma[N-M-1])}{\sin(\tilde{x}_\sigma)}
\end{aligned} \tag{B.12}$$

The determinant condition for the equations $\tilde{e}_n^{(\sigma)}$ with $n = M-1, M$ and $M+1$ not only yield these eigenvectors, but also the following transcendental equation (obtaining this equation is done absolutely analogous to Appendix A Eq. (A.13))

$$2(-1 + 2\cos(\tilde{x}_\sigma)) = \frac{\cos\left(\tilde{x}_\sigma[M - \frac{3}{2}]\right)}{\cos\left(\tilde{x}_\sigma[M - \frac{1}{2}]\right)} + \frac{\cos\left(\tilde{x}_\sigma[N - M - \frac{3}{2}]\right)}{\cos\left(\tilde{x}_\sigma[N - M - \frac{1}{2}]\right)} \tag{B.13}$$

This transcendental equation looks different from the equation of the first approach Eq. (A.15), but after the use of some trigonometric identities, their equivalence can be shown. This will be presented here in a few steps with some comments. First the separation of N (and assuming

$\sin(\tilde{x}_\sigma N) \neq 0$) is done leading to

$$\cot(\tilde{x}_\sigma N) = -\frac{\sin\left(\tilde{x}_\sigma\left[M+\frac{1}{2}\right]\right)\cos\left(\tilde{x}_\sigma\left[M-\frac{3}{2}\right]\right) + \sin\left(\tilde{x}_\sigma\left[M+\frac{3}{2}\right]\right)\cos\left(\tilde{x}_\sigma\left[M-\frac{1}{2}\right]\right)}{\text{denominator}}$$

$$-\frac{\left[4\cos(\tilde{x}_\sigma)-2\right]\sin\left(\tilde{x}_\sigma\left[M+\frac{1}{2}\right]\right)\cos\left(\tilde{x}_\sigma\left[M-\frac{1}{2}\right]\right)}{\text{denominator}} \quad \text{(B.14)}$$

$$\text{denominator} = \cos\left(\tilde{x}_\sigma\left[M+\frac{1}{2}\right]\right)\cos\left(\tilde{x}_\sigma\left[M-\frac{3}{2}\right]\right)$$
$$+\cos\left(\tilde{x}_\sigma\left[M+\frac{3}{2}\right]\right)\cos\left(\tilde{x}_\sigma\left[M-\frac{1}{2}\right]\right)$$
$$-\left[4\cos(\tilde{x}_\sigma)-2\right]\cos\left(\tilde{x}_\sigma\left[M+\frac{1}{2}\right]\right)\cos\left(\tilde{x}_\sigma\left[M-\frac{1}{2}\right]\right)$$

The next step is to further simplify the expression using the following identities

$$\cos(x)\cos(y) = \frac{1}{2}(\cos(x-y) + \cos(x+y)),$$
$$\sin(x)\cos(y) = \frac{1}{2}(\sin(x-y) + \sin(x+y)),$$
$$\cos(2x) = 2\cos^2(x) - 1 \quad \sin(2x) = 2\sin(x)\cos(x) \quad \text{(B.15)}$$

, this yields

$$\cot(\tilde{x}_\sigma N) = \frac{2\sin\left(\frac{\tilde{x}_\sigma}{2}\right)\left[2\sin(2\tilde{x}_\sigma M)\sin\left(\frac{\tilde{x}_\sigma}{2}\right) + \cos\left(\frac{\tilde{x}_\sigma}{2}\right)\right]}{2\sin^2\left(\frac{\tilde{x}_\sigma}{2}\right)\left[1 - \cos(2\tilde{x}_\sigma M)\right]}$$

Now only a few minor steps using the trigonometric identities given above have to be done to get to the same form as in Eq. (A.15)

$$\cot(\tilde{x}_\sigma N) = \cot(\tilde{x}_\sigma M) + \frac{\cot\left(\frac{\tilde{x}_\sigma}{2}\right)}{2\sin^2(\tilde{x}_\sigma M)} \quad \text{(B.16)}$$

This equation is absolute equivalent to (A.15) by just replacing the wave number q_σ used in the first approach, by the wave number used in the second approach $\tilde{x}_\sigma$. Hence this transcendental equation yields of course the exact same solutions as in the first approach.

As mentioned at the end of section "Second Method" the equivalence of the normalisation constants of the first $\mathcal{N}_{b_\sigma}$ and the second $\widetilde{\mathcal{N}}_{\tilde{b}_\sigma}$ method has to be proven. Comparing the extensive part of both normalisation constants from Eq. (A.23) and Eq. (B.12) one gets

$$\mathcal{N}_{b_\sigma}^{ext.} = \sqrt{\frac{2}{(M-1) + b_\sigma^2(N-M-1)}}$$

$$\widetilde{\mathcal{N}}_{\tilde{b}_\sigma}^{ext.} = \sqrt{\frac{2}{M-1 + \tilde{b}_\sigma^2[N-M-1]}}$$

(B.17)

To show that the extensive parts of both normalisation constants are identical one has to check whether the following equation holds

$$b_\sigma^2 \stackrel{?}{=} \tilde{b}_\sigma^2 \tag{B.18}$$

Taking the square root and using the definitions of Eq. (A.12) and Eq. (B.10) yields

$$\frac{2\sin(q_\sigma M) - \sin(q_\sigma[M-1])}{\sin(q_\sigma[N-M-1])} = \pm \frac{\cos\left(q_\sigma\left[M-\frac{1}{2}\right]\right)}{\cos\left(q_\sigma\left[N-M-\frac{1}{2}\right]\right)} \tag{B.19}$$

The "+" option is not fulfilled in general, but the "−" option yields the transcendental equation Eq. (B.16) hence showing the equivalence. The main steps to show how to achieve the transcendental equation are shown. Bringing both sides of Eq. (B.19) on a common denominator yields

$$\left(2\sin(q_\sigma M) - \sin(q_\sigma[M-1])\right)\cos\left(q_\sigma\left[N-M-\frac{1}{2}\right]\right)$$
$$+ \sin(q_\sigma[N-M-1])\cos\left(q_\sigma\left[M-\frac{1}{2}\right]\right) = 0 \tag{B.20}$$

Separation of N and assuming $\sin(q_\sigma N) \neq 0$ as before leads to

$$\cot(q_\sigma N) = \frac{Z}{N}$$
$$Z = \cos(q_\sigma[M+1])\left(\cos(q_\sigma M)\cos\left(\frac{q_\sigma}{2}\right) + \sin(q_\sigma M)\sin\left(\frac{q_\sigma}{2}\right)\right)$$
$$- \sin\left(q_\sigma\left[M+\frac{1}{2}\right]\right)\left(\sin(q_\sigma M)\cos(q_\sigma) - \cos(q_\sigma M)\sin(q_\sigma) - 2\sin(q_\sigma M)\right)$$
$$N = \sin(q_\sigma[M+1])\left(\cos(q_\sigma M)\cos\left(\frac{q_\sigma}{2}\right) + \sin(q_\sigma M)\sin\left(\frac{q_\sigma}{2}\right)\right)$$
$$+ \cos\left(q_\sigma\left[M+\frac{1}{2}\right]\right)\left(\sin(q_\sigma M)\cos(q_\sigma) - \cos(q_\sigma M)\sin(q_\sigma) - 2\sin(q_\sigma M)\right)$$

Z and N can be brought in the following form by applying trigonometric identities just like those used above

$$Z = \cos\left(\frac{q_\sigma}{2}\right) + 2\sin\left(\frac{q_\sigma}{2}\right)\sin(q_\sigma M)\cos(q_\sigma M)$$
$$N = 2\sin\left(\frac{q_\sigma}{2}\right)\sin^2(q_\sigma M) \tag{B.21}$$

With that one can immediately see the equivalence of both (extensive) normalisation constants.

Appendix C

Diagonalisation for two anharmonic bonds

The diagonalisation of the harmonic kinetic part of the Hamiltonian for $D \geq 2$

$$T_{harm} = \frac{1}{m} \sum_{\substack{k,l=1 \\ \neq M_1, M_2}}^{N-1} T_{kl}\, p_k p_l \tag{C.1}$$

with the nonzero elements of the symmetric matrix (T_{kl}) defined as

$$\begin{aligned}
T_{kk} &= \begin{cases} 2, & k=1,...,M_1-2, M_1+2,...,M_2-2, M_2+2,...,N-1 \\ \frac{3}{2}, & k=M_1-1, M_1+1, M_2-1, M_2+1 \end{cases} \\
T_{k,k+1} &= -1, \quad k=1,...,M_1-2, M_1+2,...,M_2-2, M_2+2,...,N-2 \\
T_{M_i-1,M_i+1} &= -\frac{1}{2}, \quad i=1,2
\end{aligned} \tag{C.2}$$

can be done in the standard way. The eigenvalue equation

$$\sum_{\substack{l=1 \\ l\neq M_1, M_2}}^{N-1} T_{kl} u_l^{(\sigma),\alpha} = \lambda_\sigma^{(\alpha)} u_k^{(\sigma),\alpha} \tag{C.3}$$

reads explicitly

$$(2 - \lambda_\sigma^{(\alpha)})u_k^{(\sigma),\alpha} - \left(u_{k+1}^{(\sigma),\alpha} + u_{k-1}^{(\sigma),\alpha}\right) = 0, \quad \begin{cases} 1 \leq k \leq M_1 - 2 \\ M_1 + 2 \leq k \leq M_2 - 2 \\ M_2 + 2 \leq k \leq N - 1 \end{cases} \quad (C.4)$$

$$\left(\frac{3}{2} - \lambda_\sigma^{(\alpha)}\right) u_{M_1-1}^{(\sigma),\alpha} - u_{M_1-2}^{(\sigma),\alpha} - \frac{1}{2}u_{M_1+1}^{(\sigma),\alpha} = 0, \quad k = M_1 - 1 \quad (C.5)$$

$$\left(\frac{3}{2} - \lambda_\sigma^{(\alpha)}\right) u_{M_1+1}^{(\sigma),\alpha} - u_{M_1+2}^{(\sigma),\alpha} - \frac{1}{2}u_{M_1-1}^{(\sigma),\alpha} = 0, \quad k = M_1 + 1 \quad (C.6)$$

$$\left(\frac{3}{2} - \lambda_\sigma^{(\alpha)}\right) u_{M_2-1}^{(\sigma),\alpha} - u_{M_2-2}^{(\sigma),\alpha} - \frac{1}{2}u_{M_2+1}^{(\sigma),\alpha} = 0, \quad k = M_2 - 1 \quad (C.7)$$

$$\left(\frac{3}{2} - \lambda_\sigma^{(\alpha)}\right) u_{M_2+1}^{(\sigma),\alpha} - u_{M_2+2}^{(\sigma),\alpha} - \frac{1}{2}u_{M_2-1}^{(\sigma),\alpha} = 0, \quad k = M_2 + 1 \quad (C.8)$$

The equation above is solved by the ansatz

$$u_k^{(\sigma),\alpha} = \mathcal{N}_{\sigma,\alpha} \begin{cases} \sin(q_\sigma^{(\alpha)} k), & 1 \leq k \leq M_1 - 1 \\ b_\sigma^{(s)} \cos(q_\sigma^{(s)}(M-k)), & M_1 + 1 \leq k \leq M_2 - 1 \\ b_\sigma^{(a)} \sin(q_\sigma^{(a)}(M-k)), & M_1 + 1 \leq k \leq M_2 - 1 \\ a^{(\alpha)} \sin(q_\sigma^{(\alpha)}(N-k)), & M_2 + 1 \leq k \leq N - 1 \end{cases}$$

$$a^{(\alpha)} = \begin{cases} 1, & \alpha = \text{symmetric} \\ -1, & \alpha = \text{antisymmetric} \end{cases} \quad (C.9)$$

where the parameter α can take two *values* symmetric and antisymmetric and the variable M represents "the centre" of the chain and depends on the total chain-length N being even ($M = \frac{N}{2}$) or odd ($M = \frac{N+1}{2}$). The eigenvalues can be achieved by using the ansatz and applying it to Eqs. (C.4). They are

$$\lambda_\sigma^{(\alpha)} = 2\left(1 - \cos\left(q_\sigma^{(\alpha)}\right)\right) \quad (C.10)$$

The ansatz for $k = 1, ..., M_1 - 1, M_2 + 1, ..., N - 1$ is motivated by the analogy seen from the discussion of the one anharmonic bond case (see Appendix A). Since the defects are chosen symmetric with respect to the centre of the chain M, Eq. (C.6) is equivalent to Eq. (C.7). The same counts for Eqs. (C.5) and (C.8). The ansatz for the case of $M_1 + 1 \leq k \leq M_2 - 1$ has been done using plane waves of the form

$$u_k^{(\sigma),\alpha} = A_+^{(\alpha)} e^{iq_\sigma^{(\alpha)} k} + A_-^{(\alpha)} e^{-iq_\sigma^{(\alpha)} k} \quad (C.11)$$

and following the same symmetry argument used before. That means

$\alpha =$ **symmetric**

$$\begin{aligned} u_{M_1+1}^{(\sigma),s} &\overset{!}{=} u_{M_2-1}^{(\sigma),s} \\ u_{M_1+2}^{(\sigma),s} &\overset{!}{=} u_{M_2-2}^{(\sigma),s} \\ \vdots &\overset{!}{=} \vdots \end{aligned} \qquad (C.12)$$

$\alpha =$ **antisymmetric**

$$\begin{aligned} u_{M_1+1}^{(\sigma),a} &\overset{!}{=} -u_{M_2-1}^{(\sigma),a} \\ u_{M_1+2}^{(\sigma),a} &\overset{!}{=} -u_{M_2-2}^{(\sigma),a} \\ \vdots &\overset{!}{=} \vdots \end{aligned} \qquad (C.13)$$

The coefficient is defined as

$$\begin{aligned} b_\sigma^{(s)} &= 2A_+^{(s)} e^{\frac{iq_\sigma^{(s)} N}{2}}, \quad A_-^{(s)} = A_+^{(s)} e^{iq_\sigma^{(s)} N} \\ b_\sigma^{(a)} &= 2iA_+^{(a)} e^{\frac{iq_\sigma^{(a)} N}{2}}, \quad A_-^{(a)} = -A_+^{(a)} e^{iq_\sigma^{(a)} N} \end{aligned} \qquad (C.14)$$

and its value is easily derived using the ansatz and taking Eqs. (C.5), (C.7). The procedure is analogous to the one anharmonic bond case (see Appendix A). For N being even it reads as

$$b_\sigma^{\text{even},(\alpha)} = \begin{cases} \frac{2\sin\left(q_\sigma^{(s)}[\frac{N-D}{2}]\right) - \sin\left(q_\sigma^{(s)}[\frac{N-D}{2}-1]\right)}{\cos\left(q_\sigma^{(s)}[\frac{D}{2}-1]\right)}, & \text{symmetric} \\ \frac{2\sin\left(q_\sigma^{(a)}[\frac{N-D}{2}]\right) - \sin\left(q_\sigma^{(a)}[\frac{N-D}{2}-1]\right)}{\sin\left(q_\sigma^{(a)}[\frac{D}{2}-1]\right)}, & \text{antisymmetric} \end{cases} \qquad (C.15)$$

whereas for N being odd, there is only a slight difference

$$b_\sigma^{\text{odd},(\alpha)} = \begin{cases} \frac{2\sin\left(q_\sigma^{(s)}[\frac{N-D}{2}]\right) - \sin\left(q_\sigma^{(s)}[\frac{N-D}{2}-1]\right)}{\cos\left(q_\sigma^{(s)}[\frac{D-1}{2}]\right)}, & \text{symmetric} \\ \frac{2\sin\left(q_\sigma^{(a)}[\frac{N-D}{2}]\right) - \sin\left(q_\sigma^{(a)}[\frac{N-D}{2}-1]\right)}{\sin\left(q_\sigma^{(a)}[\frac{D-1}{2}]\right)}, & \text{antisymmetric} \end{cases} \qquad (C.16)$$

Since the thermodynamic limit is taken later on, the difference between even and odd N disappear like it was discussed in the one anharmonic bond case. This is the reason, that from now on, only N being even will be discussed. For easier calculation only the parameters N, D are used from now on. All transformations necessary to achieve functions dependent only on N, D are:

$$M_1 = \frac{N-D}{2}$$

$M_2 = \frac{N+D}{2}$

The transcendental equations for the symmetric and antisymmetric case arise from Eqs. (C.6), (C.8) and the explicit calculation will be shown for the symmetric case (the antisymmetric case is done analogously). The transcendental equation in the symmetric case reads as

$$\mathcal{N}_{\sigma,s}\left[\left(2\cos\left(q_\sigma^{(s)}\right)-\frac{1}{2}\right)b_\sigma^{(s)}\cos\left(q_\sigma^{(s)}\left[\frac{D}{2}-1\right]\right)-b_\sigma^{(s)}\cos\left(q_\sigma^{(s)}\left[\frac{D}{2}-2\right]\right)\right.$$
$$\left.-\frac{1}{2}\sin\left(q_\sigma^{(s)}\left[\frac{N-D}{2}-1\right]\right)\right]=0 \qquad (C.17)$$

Using the fact, that the normalisation constant is not zero the transcendental equation can be put into the following form

$$\left[\cos\left(\frac{Dq_\sigma^{(s)}}{2}\right)-\frac{1}{2}\cos\left(q_\sigma^{(s)}\left[\frac{D}{2}-1\right]\right)\right]b_\sigma^{(s)} = \frac{1}{2}\sin\left(q_\sigma^{(s)}\left[\frac{N-D}{2}-1\right]\right)$$

and plugging in the result of Eq. (C.16) for the coefficient, the transcendental equation can be further simplified into

$$2\sin\left(q_\sigma^{(s)}\left[\frac{N-D}{2}\right]\right)\cos\left(\frac{q_\sigma^{(s)}D}{2}\right)-\sin\left(q_\sigma^{(s)}\left[\frac{N-D}{2}-1\right]\right)\cos\left(\frac{q_\sigma^{(s)}D}{2}\right)$$
$$-\sin\left(q_\sigma^{(s)}\left[\frac{N-D}{2}\right]\right)\cos\left(q_\sigma^{(s)}\left[\frac{D}{2}-1\right]\right)=0$$

Now using the following trigonometric identity $\sin(x)\cos(y) = \frac{1}{2}\left(\sin(x+y)+\sin(x-y)\right)$ the transcendental equation transforms to

$$\sin\left(q_\sigma^{(s)}\left[\frac{N}{2}-D\right]\right)+\sin\left(\frac{q_\sigma^{(s)}N}{2}\right)-\sin\left(q_\sigma^{(s)}\left[\frac{N}{2}-1\right]\right)$$
$$-\frac{1}{2}\left[\sin\left(q_\sigma^{(s)}\left[\frac{N}{2}-D-1\right]\right)+\sin\left(q_\sigma^{(s)}\left[\frac{N}{2}-D+1\right]\right)\right]=0$$

The next trigonometric identities used are $\sin(x)\pm\sin(y)=2\sin\left(\frac{x\pm y}{2}\right)\cos\left(\frac{x\mp y}{2}\right)$. With those identities the equation takes the form

$$\sin\left(q_\sigma^{(s)}\left[\frac{N}{2}-D\right]\right)\left[1-\cos\left(q_\sigma^{(s)}\right)\right]+2\cos\left(q_\sigma^{(s)}\left[\frac{N-1}{2}\right]\right)\sin\left(\frac{q_\sigma^{(s)}}{2}\right)=0$$

Since $q_\sigma^{(s)} \in (0,\pi)$ (due to the open chain) the factor $\sin\left(\frac{q_\sigma^{(s)}}{2}\right)$ is never zero and can be cancelled out by division, giving

$$\sin\left(q_\sigma^{(s)}\left[\frac{N}{2}-D\right]\right)\sin\left(\frac{q_\sigma^{(s)}}{2}\right)+\cos\left(q_\sigma^{(s)}\left[\frac{N-1}{2}\right]\right)=0$$

The discussion of this transcendental equation is more descriptive if the parameters N, D are separated. Using basic trigonometric identities one gets

$$\left[\sin\left(\frac{q_\sigma^{(s)}N}{2}\right)\cos\left(q_\sigma^{(s)}D\right) - \cos\left(\frac{q_\sigma^{(s)}N}{2}\right)\sin\left(q_\sigma^{(s)}D\right)\right]\sin\left(\frac{q_\sigma^{(s)}}{2}\right)$$
$$+\cos\left(\frac{q_\sigma^{(s)}N}{2}\right)\cos\left(\frac{q_\sigma^{(s)}}{2}\right) + \sin\left(\frac{q_\sigma^{(s)}N}{2}\right)\sin\left(\frac{q_\sigma^{(s)}}{2}\right) = 0$$

Again cancelling a factor of $\sin\left(\frac{q_\sigma^{(s)}}{2}\right)$ yields

$$\sin\left(\frac{q_\sigma^{(s)}N}{2}\right)\left[\cos\left(q_\sigma^{(s)}D\right)+1\right] = \cos\left(\frac{q_\sigma^{(s)}N}{2}\right)\left[\sin\left(q_\sigma^{(s)}D\right) - \cot\left(\frac{q_\sigma^{(s)}}{2}\right)\right]$$

Assuming $\cos\left(\frac{q_\sigma^{(s)}D}{2}\right) \neq 0$ the final forms (the antisymmetric requires $\sin\left(\frac{q_\sigma^{(a)}D}{2}\right) \neq 0$, is also presented), fit for discussion are achieved

$$\text{symmetric} \qquad \tan\left(\frac{q_\sigma^{(s)}N}{2}\right) = \underbrace{\frac{\sin(q_\sigma^{(s)}D) - \cot\left(\frac{q_\sigma^{(s)}}{2}\right)}{2\cos^2\left(\frac{q_\sigma^{(s)}D}{2}\right)}}_{f_s\left(q_\sigma^{(s)},D\right)} \qquad (C.18)$$

$$\text{antisymmetric} \qquad \cot\left(\frac{q_\sigma^{(a)}N}{2}\right) = \underbrace{\frac{\sin(q_\sigma^{(a)}D) + \cot\left(\frac{q_\sigma^{(a)}}{2}\right)}{2\sin^2\left(\frac{q_\sigma^{(a)}D}{2}\right)}}_{f_a\left(q_\sigma^{(a)},D\right)} \qquad (C.19)$$

Since both transcendental equations are not analytically solvable a detailed discussion for the approximative solutions is given. Considering the symmetric case, the left hand side (l.h.s.) shows divergences for $q_\sigma^{(s)} = \frac{2\sigma-1}{N}\pi$, $\sigma = 1, ..., \frac{N}{2}$. The form of these divergences show the following behaviour

$$\lim_{q_\sigma^{(s)}\searrow 0}\tan\left(\frac{q_\sigma^{(s)}N}{2}\right) = 0, \qquad \lim_{q_\sigma^{(s)}\nearrow \pi}\tan\left(\frac{q_\sigma^{(s)}N}{2}\right) = 0,$$

$$\lim_{q_\sigma^{(s)}\searrow \frac{2\sigma-1}{N}\pi}\tan\left(\frac{q_\sigma^{(s)}N}{2}\right) = -\infty, \qquad \lim_{q_\sigma^{(s)}\nearrow \frac{2\sigma-1}{N}\pi}\tan\left(\frac{q_\sigma^{(s)}N}{2}\right) = \infty, \qquad \sigma = 1, ..., \frac{N}{2}$$

(C.20)

which means the l.h.s. *oscillates* from $-\infty$ to ∞ in every interval $[\frac{2\sigma-1}{N}\pi, \frac{2\sigma+1}{N}\pi]$, $\sigma = 1, ..., \frac{N}{2} - 1$, except for the first interval $[0, \frac{\pi}{N}]$, where the oscillation starts at 0 and the last interval where the oscillation ends at 0.

The right hand side (r.h.s.) $f_s\left(q_\sigma^{(s)}, D\right)$ has divergences at $q_\sigma^{(s)} = \frac{2\sigma-1}{D}\pi$, $\sigma = 1,...,\frac{D}{2}$. Looking at the form of these divergences show the following behaviour

$$\lim_{q_\sigma^{(s)} \searrow 0} f_s\left(q_\sigma^{(s)}, D\right) = -\infty, \quad \lim_{q_\sigma^{(s)} \nearrow \pi} f_s\left(q_\sigma^{(s)}, D\right) = \infty,$$

$$\lim_{q_\sigma^{(s)} \searrow \frac{2\pi\sigma}{D}} f_s\left(q_\sigma^{(s)}, D\right) = -\infty, \quad \lim_{q_\sigma^{(s)} \nearrow \frac{2\pi\sigma}{D}} f_s\left(q_\sigma^{(s)}, D\right) = -\infty, \quad \sigma = 1,...,\frac{D}{2} \tag{C.21}$$

Since $D < N$ it is easy to see, that in the first interval $[0, \frac{\pi}{N}]$ and the last interval $\left[\frac{(N-1)\pi}{N}, \pi\right]$ there is no intersection of the l.h.s. and the r.h.s.. All other intervals have exactly one intersection, this reduces the total number of solutions to $\frac{N}{2} - 1$ for the symmetric case.

Regarding the case of $D = \mathcal{O}(N)$ we are allowed to choose $D = \frac{N}{2}$, since it does not matter how large D exactly is, the only important thing is, that it scales with N. In this case, we get

$$q_\sigma^{(s)} = \frac{2\sigma - 1}{N - 1}\pi \tag{C.22}$$

as exact solution.

The same discussion has to be done for the antisymmetric case. The l.h.s. has divergences at $q_\sigma^{(a)} = \frac{2\pi\sigma}{N}$, $\sigma = 1,...,\frac{N}{2}$. The form of these divergences show the following behaviour

$$\lim_{q_\sigma^{(a)} \searrow 0} \cot\left(\frac{q_\sigma^{(a)} N}{2}\right) = \infty, \quad \lim_{q_\sigma^{(a)} \nearrow \pi} \cot\left(\frac{q_\sigma^{(a)} N}{2}\right) = 0,$$

$$\lim_{q_\sigma^{(a)} \searrow \frac{2\pi\sigma}{N}} \cot\left(\frac{q_\sigma^{(a)} N}{2}\right) = \infty, \quad \lim_{q_\sigma^{(a)} \nearrow \frac{2\pi\sigma}{N}} \cot\left(\frac{q_\sigma^{(a)} N}{2}\right) = -\infty, \quad \sigma = 1,...,\frac{N}{2}-1 \tag{C.23}$$

which means the l.h.s. *oscillates* from ∞ to $-\infty$ in every interval $[\frac{2\pi\sigma}{N}, \frac{2\pi(\sigma+1)}{N}]$, $\sigma = 0,...,\frac{N}{2}-1$. The r.h.s. diverges for $q_\sigma^{(a)} = \frac{2\pi\sigma}{D}$ $\sigma = 1,...,\frac{D}{2}$. The divergences behave as follows

$$\lim_{q_\sigma^{(a)} \searrow 0} f_a\left(q_\sigma^{(a)}, D\right) = \infty, \quad \lim_{q_\sigma^{(a)} \nearrow \pi} f_a\left(q_\sigma^{(a)}, D\right) = 0,$$

$$\lim_{q_\sigma^{(a)} \searrow \frac{2\pi\sigma}{D}} f_a\left(q_\sigma^{(a)}, D\right) = \infty, \quad \lim_{q_\sigma^{(a)} \nearrow \frac{2\pi\sigma}{D}} f_a\left(q_\sigma^{(a)}, D\right) = \infty \tag{C.24}$$

Since $D < N$ it is easy to see, that in the first interval $[0, \frac{2\pi}{N}]$ and the last interval $\left[\frac{(N-1)\pi}{N}, \pi\right]$ there is no intersection of the l.h.s. and the r.h.s.. All other intervals have exactly one intersection, this reduces the total number of solutions to $\frac{N}{2} - 2$ for the antisymmetric case.

Considering $D = \mathcal{O}(N)$, again choosing $D = \frac{N}{2}$, we get the exact result

$$q_\sigma^{(a)} = \frac{2(2\sigma - 1)}{N - 2}\pi, \quad \sigma = 1,...,\frac{N}{4} - 1$$

$$q_\sigma^{(a)} = \frac{4\pi\sigma}{N}, \quad \sigma = 1,...,\frac{N}{4} - 1 \tag{C.25}$$

Note that through the choice of $D = M_2 - M_1$ and $N = M_2 + M_1$ even, $\frac{N}{4}$ must be an integer. For even N the number of symmetric and antisymmetric solutions add up to a total number of $N - 3$ solutions as it should be.

Now the case of an odd N has to be discussed. Even though the transcendental equations differ slightly from the ones achieved from even N one immediately sees, that the symmetric case and the antisymmetric case provide $\frac{N-3}{2}$ solutions each, for the same reasons as discussed before. As before adding up the symmetric and the antisymmetric case leads to $N - 3$ solutions of the transcendental equation as it should be.

In the thermodynamic limit ($N \to \infty$) both cases (N being even or odd) yield homogeneously distributed solutions $q_\sigma^{(\alpha)} \in (0, \pi)$.

That is the reason why the normalisation constant will only be calculated for N being even and hence $M = \frac{N}{2}$. Using the eigenvectors one can calculate the normalisation constant for both cases (symmetric and antisymmetric). It yields

$$1 = (\mathcal{N}_{\sigma,s})^2 \left[\sum_{k=1}^{M_1-1} \sin^2\left(q_\sigma^{(s)} k\right) + \left(b_\sigma^{(s)}\right)^2 \sum_{k=M_1+1}^{M_2-1} \cos^2\left(q_\sigma^{(s)} \left[\frac{N}{2} - k\right]\right) \right.$$
$$\left. + \sum_{k=M_2+1}^{N-1} \sin^2\left(q_\sigma^{(s)}[N-k]\right) \right]$$

and

$$1 = (\mathcal{N}_{\sigma,a})^2 \left[\sum_{k=1}^{M_1-1} \sin^2\left(q_\sigma^{(a)} k\right) + \left(b_\sigma^{(a)}\right)^2 \sum_{k=M_1+1}^{M_2-1} \sin^2\left(q_\sigma^{(a)} \left[\frac{N}{2} - k\right]\right) \right.$$
$$\left. + \sum_{k=M_2+1}^{N-1} \sin^2\left(q_\sigma^{(a)}[N-k]\right) \right] \quad (C.26)$$

The calculation is done in the same manner as in the one anharmonic bond case. Replacing M_1 by $\frac{N-D}{2}$ and M_2 by $\frac{N+D}{2}$, yields the final results

$$\mathcal{N}_{\sigma,s} = \sqrt{\frac{2}{N - D - 1 - \frac{\sin(q_\sigma^{(s)}[N-D-1])}{\sin(q_\sigma^{(s)})} + \left(b_\sigma^{(s)}\right)^2 \left[D - 1 + \frac{\sin(q_\sigma^{(s)}[D-1])}{\sin(q_\sigma^{(s)})}\right]}} \quad (C.27)$$

and

$$\mathcal{N}_{\sigma,a} = \sqrt{\frac{2}{N - D - 1 - \frac{\sin(q_\sigma^{(a)}[N-D-1])}{\sin(q_\sigma^{(a)})} + \left(b_\sigma^{(a)}\right)^2 \left[D - 1 - \frac{\sin(q_\sigma^{(a)}[D-1])}{\sin(q_\sigma^{(a)})}\right]}} \quad (C.28)$$

The low frequency limit ($q_\sigma^{(\alpha)} \ll 1$) has to be discussed in detail. Starting with the coefficient $b_\sigma^{(\alpha)}$ for the symmetric case (only the symmetric will be shown, the procedure for the antisymmetric is identical) first a separation of the arguments of Eq. (C.16) is performed

$$
\begin{aligned}
b_\sigma^{(s)} &= \frac{2\sin\left(\frac{q_\sigma^{(s)}N}{2}\right)\cos\left(\frac{q_\sigma^{(s)}D}{2}\right) - 2\cos\left(\frac{q_\sigma^{(s)}N}{2}\right)\sin\left(\frac{q_\sigma^{(s)}D}{2}\right)}{\cos\left(q_\sigma^{(s)}\left[\frac{D}{2}-1\right]\right)} \\
&\quad - \frac{\sin\left(\frac{q_\sigma^{(s)}N}{2}\right)\cos\left(q_\sigma^{(s)}\left[\frac{D}{2}+1\right]\right) - \cos\left(\frac{q_\sigma^{(s)}N}{2}\right)\sin\left(q_\sigma^{(s)}\left[\frac{D}{2}+1\right]\right)}{\cos\left(q_\sigma^{(s)}\left[\frac{D}{2}-1\right]\right)}
\end{aligned} \quad (C.29)
$$

Now the following approximations are helpful and achieved using the transcendental equations with the help of basic trigonometric identities. They are

$$
\sin\left(\frac{q_\sigma^{(s)}N}{2}\right) = \frac{f_s\left(q_\sigma^{(s)},D\right)}{\sqrt{1+f_s^2\left(q_\sigma^{(s)},D\right)}} \stackrel{q_\sigma^{(s)}\ll 1}{\approx} -1
$$

$$
\cos\left(\frac{q_\sigma^{(s)}N}{2}\right) = \frac{1}{\sqrt{1+f_s^2\left(q_\sigma^{(s)},D\right)}} \stackrel{q_\sigma^{(s)}\ll 1}{\approx} q_\sigma^{(s)}
$$

$$
\sin\left(q_\sigma^{(s)}N\right) = \frac{2f_s\left(q_\sigma^{(s)},D\right)}{1+f_s^2\left(q_\sigma^{(s)},D\right)} \stackrel{q_\sigma^{(s)}\ll 1}{\approx} -2q_\sigma^{(s)}
$$

$$
\cos\left(q_\sigma^{(s)}N\right) = \frac{1-f_s^2\left(q_\sigma^{(s)},D\right)}{1+f_s^2\left(q_\sigma^{(s)},D\right)} \stackrel{q_\sigma^{(s)}\ll 1}{\approx} -1
$$

$$
\sin\left(\frac{q_\sigma^{(a)}N}{2}\right) = \frac{1}{\sqrt{1+f_a^2\left(q_\sigma^{(a)},D\right)}} \stackrel{q_\sigma^{(a)}\ll 1}{\approx} \frac{D^2}{4}\left(q_\sigma^{(a)}\right)^3
$$

$$
\cos\left(\frac{q_\sigma^{(a)}N}{2}\right) = \frac{f_a\left(q_\sigma^{(a)},D\right)}{\sqrt{1+f_a^2\left(q_\sigma^{(a)},D\right)}} \stackrel{q_\sigma^{(a)}\ll 1}{\approx} 1
$$

$$
\sin\left(q_\sigma^{(a)}N\right) = \frac{2f_a\left(q_\sigma^{(a)},D\right)}{1+f_a^2\left(q_\sigma^{(a)},D\right)} \stackrel{q_\sigma^{(a)}\ll 1}{\approx} \frac{D^2}{2}\left(q_\sigma^{(a)}\right)^3
$$

$$
\cos\left(q_\sigma^{(a)}N\right) = \frac{f_a^2\left(q_\sigma^{(a)},D\right)-1}{1+f_a^2\left(q_\sigma^{(a)},D\right)} \stackrel{q_\sigma^{(a)}\ll 1}{\approx} 1 \quad (C.30)
$$

These approximations yield the final result for the coefficient $b_\sigma^{(s)}$ for $q_\sigma^{(s)}\ll 1$

$$
b_\sigma^{(s)} \approx -1 + \mathcal{O}(q_\sigma^{(s)}) \quad (C.31)
$$

The antisymmetric case is done analogously yielding

$$b_\sigma^{(a)} \approx -1 + \mathcal{O}(q_\sigma^{(a)}) \tag{C.32}$$

which yields for low frequencies and respectively large N

$$\mathcal{N}_{\sigma,\alpha} \sim \sqrt{\frac{2}{N}} \tag{C.33}$$

the expected N-dependence of the normalisation constant.

Appendix D

Calculation of the influence functional

Here the explicit calculation of the functional $\mathcal{F}^n$ defined in Eq. (4.70) is shown. The result is is given in Eq. (4.74) as $\mathcal{F}^n = \mathcal{F}^n_{(1)}\mathcal{F}^n_{(2)}\mathcal{G}^n_{(12)}$. The functions $L^{ab}_1(\tau), L^{ab}_2(\tau)$ are given with their dependence on the coupling constants $c_{a,\sigma}, c_{b,\sigma}$. Only here in the Appendix the full calculation with the dependence of the coupling constants for each anharmonic bond is given. Starting with the bilinear blip-term one gets the following argument by applying the transformations Eq. (4.43). The calculation of the first term (with the indices a, b for the functions $L^{ab}_2(\tau), L^{ab}_1(\tau)$ of Eq. (4.34)) with the use of the transformations Eqs. (4.43) yields:

$$\sum_{a,b=1}^{2} \int_0^t d\tau \int_0^\tau d\tau' \, L^{ab}_2(\tau-\tau')\xi^{(a)}(\tau)\xi^{(b)}(\tau') \tag{D.1}$$

$$= \sum_{a,b=1}^{2} \sum_{j,j'=1}^{n} \zeta_j^{(a)} \zeta_{j'}^{(b)} \int_0^t d\tau \int_0^\tau d\tau' \, L_2^{ab}(\tau-\tau') \Big[\Theta(\tau-t_{2j-1})\Theta(\tau'-t_{2j'-1})$$

$$- \Theta(\tau-t_{2j-1})\Theta(\tau'-t_{2j'}) - \Theta(\tau-t_{2j})\Theta(\tau'-t_{2j'-1}) + \Theta(\tau-t_{2j})\Theta(\tau'-t_{2j'}) \Big]$$

$$= \sum_{a,b=1}^{2} \sum_{j,j'=1}^{n} \zeta_j^{(a)} \zeta_{j'}^{(b)} \Bigg[\int_{t_{2j-1}}^t d\tau \left(\int_{t_{2j'-1}}^\tau d\tau' \, L_2^{ab}(\tau-\tau') - \int_{t_{2j'}}^\tau d\tau' \, L_2^{ab}(\tau-\tau') \right)$$

$$- \int_{t_{2j}}^t d\tau \left(\int_{t_{2j'}}^\tau d\tau' \, L_2^{ab}(\tau-\tau') - \int_{t_{2j'-1}}^\tau d\tau' \, L_2^{ab}(\tau-\tau') \right) \Bigg]$$

$$= \sum_{a,b=1}^{2} \sum_{j,j'=1}^{n} \zeta_j^{(a)} \zeta_{j'}^{(b)} \Bigg[\int_{t_{2j-1}}^t d\tau \left(\mathcal{L}_2^{ab}(\tau-t_{2j'-1}) - \mathcal{L}_2^{ab}(\tau-t_{2j'}) \right)$$

$$+ \int_{t_{2j}}^t d\tau \left(\mathcal{L}_2^{ab}(\tau-t_{2j'}) - \mathcal{L}_2^{ab}(\tau-t_{2j'-1}) \right) \Bigg]$$

$$= \sum_{a,b=1}^{2} \sum_{j,j'=1}^{n} \zeta_j^{(a)} \zeta_{j'}^{(b)} \Bigg[Q_2^{(ab)}(t_{2j-1}-t_{2j'}) + Q_2^{(ab)}(t_{2j}-t_{2j'-1})$$

$$- Q_2^{(ab)}(t_{2j-1}-t_{2j'-1}) - Q_2^{(ab)}(t_{2j}-t_{2j'}) \Bigg]$$

$$= \sum_{a,b=1}^{2} \sum_{j,j'=1}^{n} \zeta_j^{(a)} \zeta_{j'}^{(b)} \Lambda_{jj'}^{(ab)} \qquad (D.2)$$

The functions $Q_2^{(ab)}$ are defined in Eq. (4.75) and the function $\Lambda_{jj'}^{(ab)}$ is defined in Eq. (4.76). This is the blip-blip-interaction and self-energy part of the argument of the exponential (together with the constant factor of $\frac{q_0^2}{\pi\hbar}$) of Eq. (4.74). And now the same procedure for the term containing the function $L_1^{ab}(\tau)$ yields:

$$i \sum_{a,b=1}^{2} \int_0^t d\tau \int_0^\tau d\tau' \, L_1^{ab}(\tau-\tau') \xi^{(a)}(\tau) \chi^{(b)}(\tau') \qquad (D.3)$$

$$= i \sum_{a,b=1}^{2} \sum_{j'=0}^{n-1} \sum_{j=j'+1}^{n} \eta_{j'}^{(a)} \zeta_{j}^{(b)} \int_{0}^{t} d\tau \int_{0}^{\tau} d\tau' \, L_{1}^{ab}(\tau - \tau') \Big[\Theta(\tau - t_{2j-1}) \Theta(\tau' - t_{2j'})$$

$$- \Theta(\tau - t_{2j-1}) \Theta(\tau' - t_{2j'+1}) - \Theta(\tau - t_{2j}) \Theta(\tau' - t_{2j'}) + \Theta(\tau - t_{2j}) \Theta(\tau' - t_{2j'+1}) \Big]$$

$$= i \sum_{a,b=1}^{2} \sum_{j'=0}^{n-1} \sum_{j=j'+1}^{n} \eta_{j'}^{(a)} \zeta_{j}^{(b)} \Bigg[\int_{t_{2j-1}}^{t} d\tau \left(\int_{t_{2j'}}^{\tau} d\tau' \, L_{1}^{ab}(\tau - \tau') - \int_{t_{2j'+1}}^{\tau} d\tau' \, L_{1}^{ab}(\tau - \tau') \right)$$

$$+ \int_{t_{2j}}^{t} d\tau \left(\int_{t_{2j'+1}}^{\tau} d\tau' \, L_{1}^{ab}(\tau - \tau') - \int_{t_{2j'}}^{\tau} d\tau' \, L_{1}^{ab}(\tau - \tau') \right) \Bigg]$$

$$= i \sum_{a,b=1}^{2} \sum_{j'=0}^{n-1} \sum_{j=j'+1}^{n} \eta_{j'}^{(a)} \zeta_{j}^{(b)} \Bigg[\int_{t_{2j-1}}^{t} d\tau \left(\mathcal{L}_{1}^{ab}(\tau - t_{2j'}) - \mathcal{L}_{1}^{ab}(\tau - t_{2j'+1}) \right)$$

$$+ \int_{t_{2j}}^{t} d\tau \left(\mathcal{L}_{1}^{ab}(\tau - t_{2j'+1}) - \mathcal{L}_{1}^{ab}(\tau - t_{2j'}) \right) \Bigg]$$

$$= i \sum_{a,b=1}^{2} \sum_{j'=0}^{n-1} \sum_{j=j'+1}^{n} \eta_{j'}^{(a)} \zeta_{j}^{(b)} \Bigg[Q_{1}^{(ab)}(t_{2j-1} - t_{2j'}) + Q_{1}^{(ab)}(t_{2j} - t_{2j'+1})$$

$$- Q_{1}^{(ab)}(t_{2j-1} - t_{2j'+1}) - Q_{1}^{(ab)}(t_{2j} - t_{2j'}) \Bigg]$$

$$= i \sum_{a,b=1}^{2} \sum_{j'=0}^{n-1} \sum_{j=j'+1}^{n} \eta_{j'}^{(a)} \zeta_{j}^{(b)} X_{jj'}^{(ab)} \qquad (D.4)$$

The functions $Q_1^{(ab)}$ are defined in Eq. (4.75) and the function $X_{jj'}^{(ab)}$ is defined in Eq. (4.76). This is the blip-sojourn-interaction part of the argument of the exponential (together with the constant factor of $\frac{q_0^2}{\pi\hbar}$) of Eq. (4.74).

Appendix E

Density matrix for two anharmonic bonds

Starting with the Hamiltonian for two anharmonic bonds presented in Eq. (4.1) it can be brought, omitting the c.o.m., into the form of Eqs. (4.12), (4.13), (4.14)

$$H = H'_d + H'_{int} + H_{harm} \tag{E.1}$$

$$H'_d = \frac{1}{m}\left(p_{M_1}^2 + p_{M_2}^2\right) + \frac{C}{4}\left(q_{M_1}^2 + q_{M_2}^2\right) + V_0(q_{M_1}) + V_0(q_{M_2}) \tag{E.2}$$

$$H_{harm} = \frac{1}{2}\sum_{\sigma=1}^{N-3}\left[\lambda_\sigma P_\sigma^2 + CQ_\sigma^2\right] \tag{E.3}$$

$$H'_{int} = -\sum_{\sigma=1}^{N-3} Q_\sigma \sum_{a=1}^{2} c_{a,\sigma} q_{M_a} \tag{E.4}$$

Including the potential renormalisation of H'_d in H'_{int} allows us to rewrite the above given Hamiltonian in the form of

$$H = H_d + H_{int} + H_{harm} \tag{E.5}$$

$$H_d = \frac{1}{m}\left(p_{M_1}^2 + p_{M_2}^2\right) + V_0(q_{M_1}) + V_0(q_{M_2}) \tag{E.6}$$

$$H_{harm} = \frac{1}{2}\sum_{\sigma=1}^{N-3}\left[\lambda_\sigma P_\sigma^2 + CQ_\sigma^2\right] \tag{E.7}$$

$$H_{int} = \sum_{\sigma=1}^{N-3}\sum_{a=1}^{2}\left(-Q_\sigma c_{a,\sigma} q_{M_a} + \frac{1}{2}\frac{c_{a,\sigma}^2 q_{M_a}^2}{m_\sigma \omega_\sigma^2}\right) \tag{E.8}$$

From now on the explicit time dependence is shown, because for the calculation it is necessary to distinguish between time dependent and time independent quantities.
Using the Liouville-von-Neumann Equation for the time dependent density matrix $\rho_{tot}(t)$ of the

system-bath Hamiltonian one can write down:

$$\frac{d}{dt}\rho_{tot}(t) = -\frac{i}{\hbar}[H, \rho_{tot}(t)], \qquad \rho_{tot}(t) = e^{-\frac{i}{\hbar}Ht}\rho_{tot}(0)e^{\frac{i}{\hbar}Ht} \qquad (E.9)$$

The full density matrix element in spatial representation reads

$$\begin{aligned}\langle \{Q_\sigma^f\}, q_{M_1}^f, q_{M_2}^f | \rho_{tot}(t) | q_{M_1}'^f, q_{M_2}'^f, \{Q_\sigma'^f\}\rangle &= \int dq_{M_1}^i\, dq_{M_2}^i\, dq_{M_1}'^i\, dq_{M_2}'^i\, d\{Q_\sigma^i\}d\{Q_\sigma'^i\} \\ &\cdot \langle\{Q_\sigma^f\}, q_{M_1}^f, q_{M_2}^f | e^{-\frac{i}{\hbar}Ht} | q_{M_1}^i, q_{M_2}^i, \{Q_\sigma^i\}\rangle \\ &\cdot \langle\{Q_\sigma^i\}, q_{M_1}^i, q_{M_2}^i | \rho_{tot}(0) | q_{M_1}'^i, q_{M_2}'^i, \{Q_\sigma'^i\}\rangle \\ &\cdot \langle\{Q_\sigma'^i\}, q_{M_1}'^i, q_{M_2}'^i | e^{\frac{i}{\hbar}Ht} | q_{M_1}'^f, q_{M_2}'^f, \{Q_\sigma'^f\}\rangle\end{aligned} \qquad (E.10)$$

, where

$$\begin{aligned} q_{M_1}^i &= q_{M_1}(0),\ q_{M_2}^i = q_{M_2}(0),\ q_{M_1}'^i = q_{M_1}'(0),\ q_{M_2}'^i = q_{M_2}'(0) \\ q_{M_1}^f &= q_{M_1}(t),\ q_{M_2}^f = q_{M_2}(t),\ q_{M_1}'^f = q_{M_1}'(t),\ q_{M_2}'^f = q_{M_2}'(t) \end{aligned} \qquad (E.11)$$

and

$$\begin{aligned} \{Q_\sigma^i\} &= \{Q_\sigma(0)\},\ \{Q_\sigma'^i\} = \{Q_\sigma'(0)\} \\ \{Q_\sigma^f\} &= \{Q_\sigma(t)\},\ \{Q_\sigma'^f\} = \{Q_\sigma'(t)\} \end{aligned} \qquad (E.12)$$

By tracing out the bath degrees of freedom one obtains the reduced density matrix $\rho_{red}(t) = \text{Tr}_{bath}\rho_{tot}(t)$, which is done by setting $\{Q_\sigma'^f\} = \{Q_\sigma^f\}$.

$$\begin{aligned}\langle q_{M_1}^f, q_{M_2}^f | \rho_{red}(t) | q_{M_1}'^f, q_{M_2}'^f\rangle &= \int dq_{M_1}^i\, dq_{M_2}^i\, dq_{M_1}'^i\, dq_{M_2}'^i\, d\{Q_\sigma^i\}\,d\{Q_\sigma'^i\}\,d\{Q_\sigma^f\} \\ &\cdot \langle\{Q_\sigma^f\}, q_{M_1}^f, q_{M_2}^f | e^{-\frac{i}{\hbar}Ht} | q_{M_1}^i, q_{M_2}^i, \{Q_\sigma^i\}\rangle \\ &\cdot \langle\{Q_\sigma^i\}, q_{M_1}^i, q_{M_2}^i | \rho_{tot}(0) | q_{M_1}'^i, q_{M_2}'^i, \{Q_\sigma'^i\}\rangle \\ &\cdot \langle\{Q_\sigma'^i\}, q_{M_1}'^i, q_{M_2}'^i | e^{\frac{i}{\hbar}Ht} | q_{M_1}'^f, q_{M_2}'^f, \{Q_\sigma^f\}\rangle \end{aligned} \qquad (E.13)$$

Assuming the density matrix has factorising initial conditions[1] $\rho_{tot}(0) = \rho_{red}(0) \otimes \rho_{harm}(0)$ and knowing that the Hamiltonian (E.5) induces a classical action $S = S_d[q_{M_1}, q_{M_2}] + S_{harm}[\{Q_\sigma\}] + S_{int}[\{Q_\sigma\}q_{M_1}, q_{M_2}]$ one can rewrite the equation above in the following form

$$\begin{aligned}\langle q_{M_1}^f, q_{M_2}^f | \rho_{red}(t) | q_{M_1}'^f, q_{M_2}'^f\rangle &= \int dq_{M_1}^i\, dq_{M_2}^i\, dq_{M_1}'^i\, dq_{M_2}'^i\, \langle q_{M_1}^i, q_{M_2}^i | \rho_{red}(0) | q_{M_1}'^i, q_{M_2}'^i\rangle \\ &\cdot \int_{q_{M_1}^i}^{q_{M_1}^f} \mathcal{D}q_{M_1} \int_{q_{M_2}^i}^{q_{M_2}^f} \mathcal{D}q_{M_2} \int_{q_{M_1}'^i}^{q_{M_1}'^f} \mathcal{D}q_{M_1}' \int_{q_{M_2}'^i}^{q_{M_2}'^f} \mathcal{D}q_{M_2}'\, e^{\frac{i}{\hbar}(S_d[q_{M_1}, q_{M_2}] - S_d[q_{M_1}', q_{M_2}'])} \\ &\cdot \mathcal{F}\left[q_{M_1}, q_{M_2}, q_{M_1}', q_{M_2}'\right]\end{aligned} \qquad (E.14)$$

---
[1] as assumed in [21]

The first term before the path integrals describes the preparation of the initial state of the anharmonic bonds.

$$
\begin{aligned}
\mathcal{F}\left[q_{M_1}, q_{M_2}, q'_{M_1}, q'_{M_2}\right] = & \int d\{Q^i_\sigma\} d\{Q'^i_\sigma\} d\{Q^f_\sigma\} \langle \{Q^i_\sigma\}|\rho_{harm}(0)|\{Q'^i_\sigma\}\rangle \\
& \cdot \int_{\{Q^i_\sigma\}}^{\{Q^f_\sigma\}} \mathcal{D}\{Q_\sigma\} \int_{\{Q'^i_\sigma\}}^{\{Q^f_\sigma\}} \mathcal{D}\{Q'_\sigma\} \\
& \cdot e^{\frac{i}{\hbar}\left(S_{harm}[\{Q_\sigma\}] + S_{int}[\{Q_\sigma\}, q_{M_1}, q_{M_2}] - S_{harm}[\{Q'_\sigma\}] - S_{int}[\{Q'_\sigma\}, q'_{M_1}, q'_{M_2}]\right)}
\end{aligned}
$$
(E.15)

The term presented above Eq. (E.15) is called influence functional in literature. The exponential right before the influence functional containing S_d gives the bare tunnelling amplitudes of the anharmonic bonds $A[q_{M_1}]$, $A^*[q'_{M_1}]$, $B[q_{M_2}]$, $B^*[q'_{M_2}]$. The separation of the harmonic and anharmonic degrees of freedom yields a functional containing only the anharmonic bonds S_d, which describes tunnelling of both bonds without coupling to the harmonic bath. The dissipative influence of the harmonic bath is fully captured in the influence functional Eq. (E.15). The density matrix elements of the bath can be written [16, 21] as a product of all $N-3$ harmonic modes.

$$
\begin{aligned}
\langle \{Q^i_\sigma\}|\rho_{harm}(0)|\{Q'^i_\sigma\}\rangle = & \prod_{\sigma=1}^{N-3} \frac{1}{2\sinh\left(\frac{\omega_\sigma \hbar \beta}{2}\right)} \sqrt{\frac{m_\sigma \omega_\sigma}{2\pi\hbar \sinh(\omega_\sigma \hbar \beta)}} \\
& \cdot \exp\left[-\frac{m_\sigma \omega_\sigma}{2\hbar \sinh(\omega_\sigma \hbar \beta)}\left(\left[(Q^i_\sigma)^2 + (Q'^i_\sigma)^2\right]\cosh(\omega_\sigma \hbar \beta) - 2 Q^i_\sigma Q'^i_\sigma\right)\right]
\end{aligned}
$$
(E.16)

The expression for the second part of the influence functional (only the functional for $\{Q_\sigma\}, q_{M_1}, q_{M_2}$ is given, the functional for $\{Q'_\sigma\}, q'_{M_1}, q'_{M_2}$ is performed in the same way) reads after performing the path integration over the harmonic degrees of freedom [16, 21]

$$
\int_{\{Q^i_\sigma\}}^{\{Q^f_\sigma\}} \mathcal{D}\{Q_\sigma\} \, e^{\frac{i}{\hbar}\left(S_{harm}[\{Q_\sigma\}] + S_{int}[\{Q_\sigma\}, q_{M_1}, q_{M_2}]\right)} = \prod_{\sigma=1}^{N-3} \sqrt{\frac{m_\sigma \omega_\sigma}{2i\pi\hbar \sin(\omega_\sigma t)}} \, e^{\frac{i}{\hbar}\phi[q_{M_1}, q_{M_2}, Q^i_\sigma, Q^f_\sigma]}
$$

$$\phi[q_{M_1}, q_{M_2}, Q^i_\sigma, Q^f_\sigma] = \frac{m_\sigma \omega_\sigma}{2\sin(\omega_\sigma t)} \left[\left([(Q^i_\sigma)^2 + (Q^f_\sigma)^2] \cos(\omega_\sigma t) - 2Q^i_\sigma Q^f_\sigma \right) \right]$$

$$+ \sum_{a=1}^{2} \frac{Q^i_\sigma c_{a,\sigma}}{\sin(\omega_\sigma t)} \int_0^t d\tau \, \sin(\omega_\sigma(t-\tau)) \, q_{M_a}(\tau)$$

$$+ \sum_{a=1}^{2} \frac{Q^f_\sigma c_{a,\sigma}}{\sin(\omega_\sigma t)} \int_0^t d\tau \, \sin(\omega_\sigma \tau) \, q_{M_a}(\tau) - \sum_{a=1}^{2} \frac{c_{a,\sigma}^2}{2m_\sigma \omega_\sigma^2} \int_0^t d\tau \, q_{M_a}^2(\tau)$$

$$- \sum_{a,b=1}^{2} \frac{c_{a,\sigma} c_{b,\sigma}}{m_\sigma \omega_\sigma \sin(\omega_\sigma t)} \int_0^t d\tau \int_0^\tau d\tau'$$

$$\cdot \sin(\omega_\sigma(t-\tau)) \sin(\omega_\sigma \tau') \, q_{M_a}(\tau) q_{M_b}(\tau') \quad \text{(E.17)}$$

Applying the calculations given above the influence functional can be given in the Feynman-Vernon [21] form (after performing the Gaussian-Integration over $d\{Q^i_\sigma\}, d\{Q'^i_\sigma\}, d\{Q^f_\sigma\}$), not for one but two anharmonic bonds and a suitable choice of initial and final conditions for the anharmonic bonds.

$$\mathcal{F}[q_{M_1}, q_{M_2}, q'_{M_1}, q'_{M_2}] = \exp\left[-\frac{1}{\hbar} \sum_{a,b=1}^{2} \int_0^t d\tau \int_0^\tau d\tau' \left(q_{M_a}(\tau) - q'_{M_a}(\tau) \right) \right.$$

$$\left. \cdot \left(L_{ab}(\tau - \tau') q_{M_b}(\tau') - L^*_{ab}(\tau - \tau') q'_{M_b}(\tau') \right) \right] \quad \text{(E.18)}$$

where the function $L_{ab}(\tau)$ is defined as

$$L_{ab}(\tau) = \sum_{\sigma=1}^{N-3} \frac{c_{a,\sigma} c_{b,\sigma}}{2m_\sigma \omega_\sigma} \left[\coth\left(\frac{\omega_\sigma \hbar \beta}{2} \right) \cos(\omega_\sigma \tau) - i\sin(\omega_\sigma \tau) \right] \quad \text{(E.19)}$$

the influence functional presented here has the same form as in [22] with the difference, that now not only one anharmonic bond, but two are considered. The complete expression for the probability to have arrived in the final state at time t (i.e. the element $\langle q^f_{M_1}, q^f_{M_2} | \rho_{red}(t) | q^f_{M_1}, q^f_{M_2} \rangle$) of the reduced density matrix), is

$$p(t) = \int \mathcal{D} q_{M_1}(\tau) \, \mathcal{D} q'_{M_1}(\tau') \mathcal{D} q_{M_2}(\tau) \, \mathcal{D} q'_{M_2}(\tau') A[q_{M_1}] A^*[q'_{M_1}] B[q_{M_2}] B^*[q'_{M_2}]$$

$$\cdot \mathcal{F}[q_{M_1}(\tau), q'_{M_1}(\tau'); q_{M_2}(\tau), q'_{M_2}(\tau')] \quad \text{(E.20)}$$

, where $p_a(t) = \sum_i \langle a_i | \rho_{red}(t) | a_i \rangle$. Splitting the function $L_{ab}(\tau)$ into imaginary and real part $L_{ab}(\tau) = L_2^{ab}(\tau) - iL_1^{ab}(\tau)$ one gets the form needed to understand the derivation of (4.62), (4.63) in subsection "Tunnelling expectation value using extended NIBA".

Appendix F

Blip- and Sojourn charge summation

The summation of blip- and sojourn-charges for the functionals $\mathcal{F}_{(1),\text{NIBA}}^n \cdot \mathcal{F}_{(2),\text{NIBA}}^n \cdot \mathcal{G}_{\text{NIBA}}^n$, defined in Eq. (4.80) is performed here in detail. At first the summation for $P_1(t)$ is performed. Since it is obviously irrelevant if the initial state is AA or DD, as long as the final state is fixed to be the same as the initial state (for $P_1(t)$). Beginning with the summation of sojourn-charges $\{\eta_j^{(1)}\}, \{\eta_j^{(2)}\}$, $n > j \geq 1$, yields

$$\sum_{\{\eta_j^{(1)}\},\{\eta_j^{(2)}\}} \mathcal{F}_{(1),\text{NIBA}}^n \cdot \mathcal{F}_{(2),\text{NIBA}}^n \cdot \mathcal{G}_{\text{NIBA}}^n$$

$$= \text{SE}_1 \cdot \text{SE}_2 \cdot \text{BB}_{12} \cdot \underbrace{e^{\pm \frac{iq_0^2}{\pi\hbar}\left(\zeta_1^{(1)}\{Q_1(t_2-t_1)+Q_1^{(12)}(t_2-t_1)\}+\zeta_1^{(2)}\{Q_1(t_2-t_1)+Q_1^{(12)}(t_2-t_1)\}\right)}}_{f_0^+(\eta_0^{(1)}=\eta_0^{(2)}=\pm 1,\,\zeta_1^{(1)},\zeta_1^{(2)})}$$

$$\cdot \underbrace{2^{n-1}\prod_{j=2}^n \cos\left(\frac{q_0^2}{\pi\hbar}\left[\zeta_j^{(1)}\{Q_1(t_{2j}-t_{2j-1})+Q_1^{(12)}(t_{2j}-t_{2j-1})\}+\zeta_j^{(2)}\{Q_1(t_{2j}-t_{2j-1})+Q_1^{(12)}(t_{2j}-t_{2j-1})\}\right]\right)}_{g^+(\zeta_j^{(1)},\zeta_j^{(2)})}$$

The expressions $\text{SE}_1, \text{SE}_2, \text{BS}_1, \text{BS}_2, \text{BS}_{12}, \text{BB}_{12}$ are defined in Eq. (4.80). To understand the summation more easily, the restrictions applied here can be expressed mathematically as $\eta_j^{(1)} = \eta_j^{(2)}$. Now the summation of the blip-charges can be performed.

As in the sojourn-charge summation, the charges of both anharmonic bonds have to be equal ($\zeta_j^{(1)} = \zeta_j^{(2)}$), yielding

$$\text{SE}_1 \cdot \text{SE}_2 \cdot \sum_{\{\zeta_j^{(1)}\},\{\zeta_j^{(2)}\}} f_0^+(\zeta_1^{(1)},\zeta_1^{(2)}) \cdot g^+(\zeta_j^{(1)},\zeta_j^{(2)}) \cdot \text{BB}_{12}$$

$$= \text{SE}_1 \cdot \text{SE}_2 \cdot e^{-\frac{2q_0^2}{\pi\hbar}\sum_{j=1}^n Q_2^{(12)}(t_{2j}-t_{2j-1})} \cdot 2\cos\left(\frac{2q_0^2}{\pi\hbar}\left[Q_1(t_2-t_1)+Q_1^{(12)}(t_2-t_1)\right]\right)$$

$$\cdot 2^{2(n-1)}\prod_{j=2}^n \cos\left(\frac{2q_0^2}{\pi\hbar}\left[Q_1(t_{2j}-t_{2j-1})+Q_1^{(12)}(t_{2j}-t_{2j-1})\right]\right)$$

Now using the definition of Eqs. (4.86), we are able to write the result in a simpler way

$$F_n^{(1)}(\{t_{2n}\}) = 2^{2n-1} \prod_{j=1}^{n} \cos\left(\frac{2q_0^2}{\pi\hbar} Q_1^+(t_{2j} - t_{2j-1})\right) \cdot e^{-\frac{2q_0^2}{\pi\hbar} Q_2^+(t_{2j} - t_{2j-1})} \quad \text{(F.1)}$$

Performing the summation of the blip- and sojourn-charges for $P_2(t)$, we can choose an initial state of AD or DA. As before, the choice of the initial state fixes the final state. That allows a simplification of the summation as before to $\eta_j^{(1)} = -\eta_j^{(2)}$ and $\zeta_j^{(1)} = -\zeta_j^{(2)}$, yielding

$$\sum_{\{\eta_j^{(1)}\},\{\eta_j^{(2)}\}} \mathcal{F}_{(1),\text{NIBA}}^n \cdot \mathcal{F}_{(2),\text{NIBA}}^n \cdot \mathcal{G}_{\text{NIBA}}^n$$

$$= \text{SE}_1 \cdot \text{SE}_2 \cdot \text{BB}_{12} \cdot \underbrace{e^{\pm\frac{iq_0^2}{\pi\hbar}\left(\zeta_1^{(1)}\{Q_1(t_2-t_1) - Q_1^{(12)}(t_2-t_1)\} - \zeta_1^{(2)}\{Q_1(t_2-t_1) - Q_1^{(12)}(t_2-t_1)\}\right)}}_{f_0^-(\eta_0^{(1)} = -\eta_0^{(2)} = \pm 1, \zeta_1^{(1)}, \zeta_1^{(2)})}$$

$$\cdot \underbrace{2^{n-1} \prod_{j=2}^{n} \cos\left(\frac{q_0^2}{\pi\hbar}\left[\zeta_j^{(1)}\{Q_1(t_{2j}-t_{2j-1}) - Q_1^{(12)}(t_{2j}-t_{2j-1})\} - \zeta_j^{(2)}\{Q_1(t_{2j}-t_{2j-1}) - Q_1^{(12)}(t_{2j}-t_{2j-1})\}\right]\right)}_{g^-(\zeta_j^{(1)}, \zeta_j^{(2)})}$$

Now the summation of the blip-charges can be performed.
As in the sojourn-charge summation, the charges of both anharmonic bonds are not allowed to be equal, so that $\zeta_j^{(1)} = -\zeta_j^{(2)}$, yielding

$$\text{SE}_1 \cdot \text{SE}_2 \cdot \sum_{\{\zeta_j^{(1)}\},\{\zeta_j^{(2)}\}} f_0^-(\zeta_1^{(1)}, \zeta_1^{(2)}) \cdot g^-(\zeta_j^{(1)}, \zeta_j^{(2)}) \cdot \text{BB}_{12}$$

$$= \text{SE}_1 \cdot \text{SE}_2 \cdot e^{\frac{2q_0^2}{\pi\hbar} \sum_{j=1}^{n} Q_2^{(12)}(t_{2j}-t_{2j-1})} \cdot 2\cos\left(\frac{2q_0^2}{\pi\hbar}\left[Q_1(t_2-t_1) - Q_1^{(12)}(t_2-t_1)\right]\right)$$

$$\cdot 2^{2(n-1)} \prod_{j=2}^{n} \cos\left(\frac{2q_0^2}{\pi\hbar}\left[Q_1(t_{2j}-t_{2j-1}) - Q_1^{(12)}(t_{2j}-t_{2j-1})\right]\right)$$

Now using the definition of Eqs. (4.86), we are able to write the result in a simpler way

$$F_n^{(2)}(\{t_{2n}\}) = 2^{2n-1} \prod_{j=1}^{n} \cos\left(\frac{2q_0^2}{\pi\hbar} Q_1^-(t_{2j} - t_{2j-1})\right) \cdot e^{-\frac{2q_0^2}{\pi\hbar} Q_2^-(t_{2j} - t_{2j-1})} \quad \text{(F.2)}$$

Bibliography

[1] E. Joos, H. D. Zeh, C. Kiefer, D. Guilini, J. Kupsch, I.O. Stamatescu: Decoherence and the Appearance of a Classical World in Quantum Theory
Springer 2003

[2] H. Everett, Rev. Mod. Phys. **29**, 454 (1957)

[3] P. W. Anderson, B. I. Halperin, C. M. Varma, Philos. Mag. **25**, 1 (1972)

[4] W. A. Phillips, J. Low Temp. Phys. **7**, 351 (1972)

[5] C.C. Yu and A.J. Leggett, Comments Cond. Mat. Phys. **14**, 4 pp. 231-251 (1988)

[6] A.L. Burin, Yu. Kagan, L. A. Maksimov and I. Ya. Polishchuk, Phys. Rev. Lett. **80**, 13 (1998)

[7] A. L. Burin, I. Ya. Polishchuk, Journal of Low Temperature Physics, Vol. **137**, 189 (2004)

[8] A. A. Belavin, A.M. Polyakov, A.S. Schwartz, Yu. S. Tyupkin, Phys. Lett. **59B**, 1 (1975)

[9] G. 't Hooft, Phys. Rev. Lett. **37**, 8-11 (1976)

[10] S. Weinberg, Phys. Rev. Lett. **40**, 223-226 (1978)

[11] C. G. Callan, R. Dashen, D. J. Gross, Phys. Rev. D **17**, 2717-2763 (1978)

[12] R. Rajaraman, Solitons and Instantons: An Introduction to Solitons and Instantons in Quantum Field Theory
North-Holland Personal Library (1982)

[13] B. Sakita, Quantum Theory of many-variable systems and fields
World Scientific Lecture Notes in Physics Vol. 1 (1985)

[14] A. J. Leggett, S. Chakravarty, A. T. Dorsey, M. P. A. Fisher, A. Garg and W. Zwerger, Rev. Mod. Phys. **59**, 1 (1987)

[15] U. Weiss, H. Grabert, S. Linkewitz, Journal of Low Temperature Physics, **68**, 213- 244 (1987)

[16] U. Weiss, "Quantum Dissipative Systems", World Scientific Publishing Company, 2. edition (1999)

[17] H. Wipf, D. Steinbinder, K. Neumaier, P. Gutsmiedl, A. Magerl, A. J. Dianoux, Europhys. Lett. **4**, 1379 (1989)

[18] G.M. Luke et al., Phys. Rev. B **43**, 3284 (1991)

[19] K. Chun, N. O. Birge, Phys. Rev. B **48**, 11500 (1993)

[20] B. Golding, N. M. Zimmermann, S. N. Coppersmith, Phys. Rev. Lett. **68**, 998 (1992)

[21] R. P. Feynman and F. L. Vernon, Ann. Phys. (N. Y.) 24, 118 (1963)

[22] A. O. Caldeira and A. J. Leggett, Phys. Rev. Lett. **46**, 211 (1981)

[23] D.J. Scalapino, The Theory of Josephson Tunneling, in Tunneling Phenomena in Solids, eds. E. Burstein and S. Lundqvist (Plenum Press, New York, 1969) pp. 477-518

[24] W. den Boer and R. de Bruyn Ouboter, Physica B/C **98**, 185 (1980)

[25] R. J. Prance et al., Nature **289**, 543-549 (1981)

[26] S. Chakravarty, Phys. Rev. Lett. **49**, 681 (1982); A. J. Bray and M. A. Moore, Phys. Rev. Lett. **49**, 1545 (1982)

[27] E. M. Chudnovsky, Phys. Rev. B **54**, 5777 (1996)

[28] H. Grabert, P. Schramm and G.-L. Ingold, Phys. Rep. **168**, 115 (1988)

[29] M. Dubé and P. C. E. Stamp, International Journal of Modern Physics B **12**, 11 1191-1245 (1998)

[30] R. Schilling, Theoretische Physik 2 (Allgemeine Mechanik), (2007)

[31] H. A. Kramers, Physica **7**, 4 (1940)

[32] J. Kurkijärvi, Phys. Rev. B **6**, 3 (1972)

[33] A. J. Leggett, Phys. Rev. B **30**, 1208 (1984)

[34] R. Schilling, Theoretische Physik 5 (Thermodynamik und Statistische Mechanik), (2008)

[35] F. Sols and P. Bhattacharyya, Phys. Rev. B **38**, 17 (1988)

[36] V. Fleurov, R. Schilling, B. Bayani, Phys. Rev. B **78**, 184301 (2008)

[37] D. J. Thouless, Phys. Rev. **187**, 2 (1969)

i want morebooks!

Buy your books fast and straightforward online - at one of world's fastest growing online book stores! Environmentally sound due to Print-on-Demand technologies.

Buy your books online at
www.get-morebooks.com

Kaufen Sie Ihre Bücher schnell und unkompliziert online – auf einer der am schnellsten wachsenden Buchhandelsplattformen weltweit! Dank Print-On-Demand umwelt- und ressourcenschonend produziert.

Bücher schneller online kaufen
www.morebooks.de

VDM Verlagsservicegesellschaft mbH
Heinrich-Böcking-Str. 6-8   Telefon: +49 681 3720 174   info@vdm-vsg.de
D - 66121 Saarbrücken   Telefax: +49 681 3720 1749   www.vdm-vsg.de

Printed by Books on Demand GmbH, Norderstedt / Germany